Choosing, Using, and Repairing
Binoculars

By J.W. Seyfried

University Optics Inc, Ann Arbor , Michigan

Published by: University Optics Inc
P.O. Box 1205
Ann Arbor, Michigan 48106 U.S.A.

ISBN 0-934639-01-9 : $19.95

Printed in the United States of America
10 9 8 7 6 5 4 3 2

Contents

i

Introduction

It's probably a rare household that doesn't have one or two pair of binoculars knocking around somewhere. The chances are also good that they may not be in tip top condition anymore. Remember the time they fell off the porch railing? Or the time the kids took a pair camping and got them soaking wet? Now when you try and use them you get kind of a queasy feeling that things are not quite as clear as they used to be.

It wasn't many years ago you could just throw your old pair

away and pick up another inexpensive pair at a discount store. Well, times have changed! Today, if you're going shopping for binoculars, be prepared for sticker shock. Those $20 or $30 glasses of just a few years ago have suddenly become $100 or more!

What can be done? Take them to a professional optical repairman who charges $40 per hour and wait two or three months to get them back? Throw them away? Repair them yourself?

Wait a minute, isn't fixing an optical instrument way beyond beyond most people's ability? Don't you need to go to some sort of technical school to learn about them?

Optical repair is not a complex or difficult task, and that's the truth. You don't need expensive tools. You don't need to use complicated math! You will need to develop neat and careful work habits. If you've done anything like building model airplanes or trains or tuned-up the family car, then you possess many of the skills and patience needed.

The subject of binocular repair is presented here in a logical step-by-step manner. Some of the things you will learn and the skills you develop may be of use to you in other fields of interest. While I have avoided mentioning specific brand names, the basic concepts and problems covered are common to all binoculars. You would be quite surprised how little basic binocular design has changed since the 1930's!

When it comes time to replace those old glasses, the general scope of this book has been expanded to include all of the latest information regarding how to select the proper binocular to meet your particular needs as well as some tips on the correct way to use and care for it. The last few years have seen the arrival on the market of many new and useful accessories that really increase the overall enjoyment of your instrument.

Some of you may even be able to supplement your income by repairing and reselling binoculars picked up at garage or estate sales. There are still a number of WW-II German units showing up around the country that can be turned into valuable collector's items with a little careful cleaning.

CHAPTER 1

Common Myths about Binoculars *Exposed*

1) If you purchase "military specification" binoculars, you're sure of a quality product.

The question to ask here is whose military? If we're talking about binoculars built to U.S. Military specifications, they are still quite rigid. In general, the U.S. military still views binoculars as pieces of serviceable equipment. This means they are built to be taken apart and repaired. For example, the objective lenses will be mounted in a cell with precision retaining rings making it easy to adjust the collimation. One brand name that comes to mind that makes excellent U.S. Military specification units is Fujinon.

On the other hand, some European armies regard binoculars as disposable items. Remember the old U.S. Military Benrus watch? You use it till it breaks and then throw it away! Disposable binoculars may have plastic parts, lenses glued in without collimation provisions, and can not be taken apart for service.

Caution is also advised on military equipment arriving from former communist counties in Eastern Europe.

2) Zoom binoculars are the answer to most people's requirements. They give you six or seven binoculars rolled into one.

Zoom binoculars show up at repair stations on a regular basis in greater numbers in proportion to the number of units sold than any other model. They are a binocular repairman's nightmare. They are heavy. They are extremely complex. With lens systems moving within other lens systems, they are impossible to bring into or hold collimation. They are 15% to 20% heavier than conventional models. The zoom mechanism is subject to constant jams. They usually command a 20% to

25% higher price than regular models. It is impossible to seal them against moisture. They were developed as a "gimmick" to stimulate sales, the idea coming from the success enjoyed by zoom camera lenses. If someone wants to give you a pair free, politely refuse.

3) On Japanese binoculars I noticed a small gold label marked "PASSED" along with numbers 1, 2, 3, 4, etc. The salesman told me "PASSED "and the number 1 was the highest inspection grade with 3 or 4 being lesser quality grades.

The "PASSED" inspection label was developed for the export market during the 1950's as part of their quality control program. This indicates a certain number of units were inspected for specific points depending on lot size. This does not mean 100% inspection.

The city of Tokyo is divided into some six or seven inspection districts. Binocular assemblers within a district bring their products into to a district station. Number 2 or 3 on an inspection label means they were inspected at that particular station and it has no relation to quality.

4) The more powerful the binocular, the more you see !

This always comes up when boaters ask "I want to see another boat or a marking buoy four or five miles away. How far can I see with a pair of binoculars?" Well, if you think about it, you can see a long way with binoculars. You can see the moon and stars! Astronomical objects are millions of miles away, yet are readily visible.

Yet viewing on land presents its own problems. While it's true the image you are viewing comes closer as the power is increased, it doesn't follow true that you "see" more. Several other factors also come into play. A study of physics relating to light waves and lenses also tells us that with the same size lens, as power is increased, less light reaches our eyes and the image appears darker. Heat waves from the earth's surface cause the image to dance and shimmer greatly reducing image quality. As power increases, the field of view becomes narrower so we see a smaller slice of the pie, so to speak. As we near 15 or 20X, every little movement of our body, even our heart beat , shows itself in a dancing image. Beyond a certain point, usually 8 to 10X, a high-power binocular *shows us less when used* to view objects on land.

5) Quality-wise, you are always safe buying well known European brands.

Here is an interesting story! On a personal visit to a binocular factory in Korea, a lot of several hundred binoculars were being prepared for shipment. But they had no markings of any kind. What goes on here I ask? It seems that Korea had no export law requiring country-of-origin be marked on the units. They were bound for Germany. Germany in turn had no law requiring country-of-origin on binoculars being imported. After shipment to Germany, the glasses were marked with the German importer's name and the word Germany. Do you think the salesman behind the counter of your camera store knew or cared where these binoculars really came from?

6) Roof-prisms are superior to conventional porro prism binoculars.

Yes, roof-prism binoculars are compact and light weight. But during manufacture they require precise production techniques. The problem is the small size, cut, and polish required on the 90 degree angle of the prism's apex. If this angle is not exactly 90-degrees, image sharpness suffers. I would say it is 10 times harder to make a good roof prism compared to a conventional porro prism. Because of the unique shape of the roof prism, light passing through them tends to shift out of phase. If these prisms don't receive a special "phase shift" coating, they will never reveal sharp brilliant images.

To boil it all down, an excellent roof-prism binocular can equal, but never exceed an excellent quality porro prism binocular. The catch is, all that extra care that has to go into the roof prism is going to cost you a lot of money.

7) Center-focus binoculars are better than individual eye focus units.

Production figures today indicate upwards of 95% of all binoculars sold today are of the center-focus design. This doesn't mean they are superior, just that they are popular! The basic idea behind the center-focus design was they were easier for several people to use. Quickly focus with the center-focus wheel, pass them to your friend and he could refocus in an instant.

In real terms, the center- focus design is more fragile, easily damaged, impossible to waterproof, and does not lend itself to holding collimation as the eyepiece slides in and out. On the other hand, all waterproof and military designs employ the individual eye focus design. If you want a more trouble-free glass that will hold collimation during rough handling, always choose the "unpopular" individual focus model.

8) *You can easily tell if a binocular is out of alignment just by looking through it for a few minutes!*

Most of the time, the only way to tell if a binocular is out of alignment or "collimation" is to put it on some kind of test device. People have what is termed "binocular vision" and this form of vision does its best to correct defects inthe poorly collimated system. So while everything appears clear and sharp through your binocular, unknown to you, the muscles in your eye may be undergoing severe stress. The final result could be fatigue and headaches.

There are a few consumer tests that can be performed quickly when shopping for binoculars. More about them later. Just remember, unless the misalignment is extremely severe, you will never know it until you get it back home and spend a few hours with it.

9) It's now possible to buy surplus binoculars from the former Soviet Union. They sound like a good deal!

Tons of military equipment from the old Eastern Block armies are showing up now in the States. Some of it is pretty decent, a lot of it is questionable. You may be lucky enough to get some from the old Zeiss factory when it was run by the East Germans. On the other hand, units made in some of the other countries don't meet the demanding American consumer standards.

One of the big problems on most of these units we have examined is internal dirt and dust. It seems they really never grasp the clean room concept so necessary during manufacture and assembly. A lot of the parts seem to be almost hand made.

They don't fit other units of the same model and production lot. If you buy any of this stuff, just make sure the dealer has a liberal return policy; you may need to try a couple of times to get a good pair.

CHAPTER 2

General Terms

When we start the study of any new subject whether it's auto repair, building model airplanes, or fixing binoculars, we have to learn something about how it works, what the parts are called, and how the parts relate to each other. In other words, we need to know its ***nomenclature.***

A binocular is an optical and mechanical aid used to magnify distant objects so that they appear nearer to the viewer. The very word ***binocular*** means the use of both eyes. Since man is blessed with two eyes, the use of a properly functioning instrument using the binocular principle is quite natural and comfortable. In effect, we are taking two small telescopes and joining them at the center with a hinge. Modern binoculars must also contain several prisms to correctly orient the image. If you have ever looked through an astronomical telescope at the moon, you've discovered the images are upside down. Certainly not something you want while trying to follow your favorite football team!

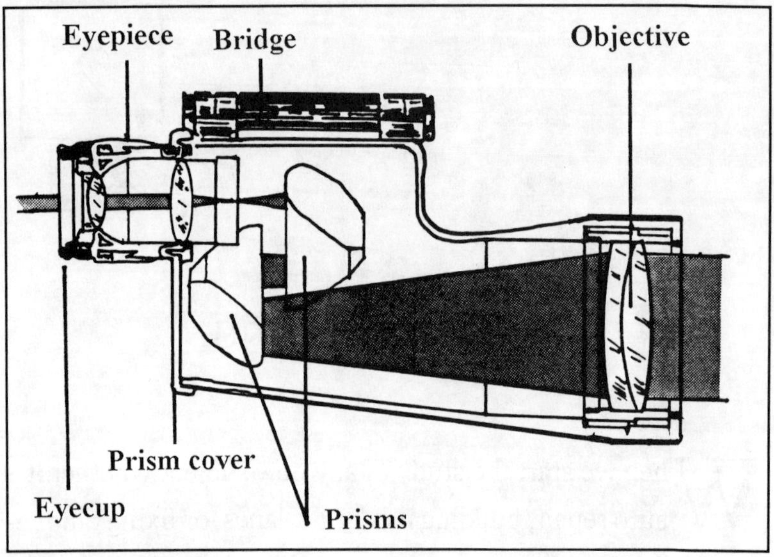

Fig. 2-1 The internal parts of a prismatic binocular.

Figure 2-1 shows a cutaway of a typical ***prismatic binocular.*** We've traced the light path completely through the instrument to show you the complexity of the system as well as the rather large number of individual glass elements we have to deal with when attempting any repairs.

Let's start with the ***eyepieces***. (Fig.2-2) The eyepieces are closest to your eyes. There are usually some type of ***eyecups*** made of plastic or rubber. These keep your eye back a bit from the glass portion of the eyepiece. They help you position your eye so you can see the entire field and they also keep your eyelashes from brushing directly on the glass lens. If they are made from rubber, many times they will fold down so people who wear glasses can get in a little closer to see the field of view.

If your binocular has a ***center focus*** design, the right eyepiece will turn or screw back and forth. This is called the "right diopter" adjustment and allows us to focus both eyes on the target. Note the + and (-) markings. If your binocular is of the ***individual eyefocus*** design, both eyepieces will be of the threaded focus design.

Center focus eyepieces will have the eyepieces tied together with an eyepiece ***bridge.*** This bridge connects the eyepieces to the central focus mechanism.

Moving on to the ***prism covers*** or end plates, we find a number of markings. These are sort of a directory of information and let us know some important things about the instrument. The markings on binoculars always tend to confuse people. The ***power*** or magnification is always expressed as a whole number followed by an "X" A binocular with a power of 7 means the object being observed looks 7 times closer compared to the same object viewed with the unaided naked eye. The next number following the X is the size or diameter of the objective or main light gathering lens of the instrument. If we have 7 X 50, the 50 is the diameter of the objective lens measured in millimeters. We always express this by saying 7 "by" 50.

The cover plates will also tell us the ***field of view*** either in degrees or feet at 1000 yards.

There will also be information on how the binocular's lenses are coated. Optical coatings increase light transmission through the system and yield brighter images. It should say ***Fully Coated, Multi-Coated, or Fully Multi-Coated*** .

If the binocular uses premium quality prisms, BAK-4 being one example, that will also be indicated.

Moving on down the instrument we come to the two body halves joined by the hinge and focusing device. There will often be some kind of wheel or "rocker" device to move the eyepieces back and forth to get sharp focus. The center focus design must also have an individual eye focus device on the right eyepiece. There is a small scale with + and - numbers to help you find and always return to the correct setting. On some military or waterproof binoculars, each individual eyepiece will have to be turned so each eyepiece will have diopter + or - markings.

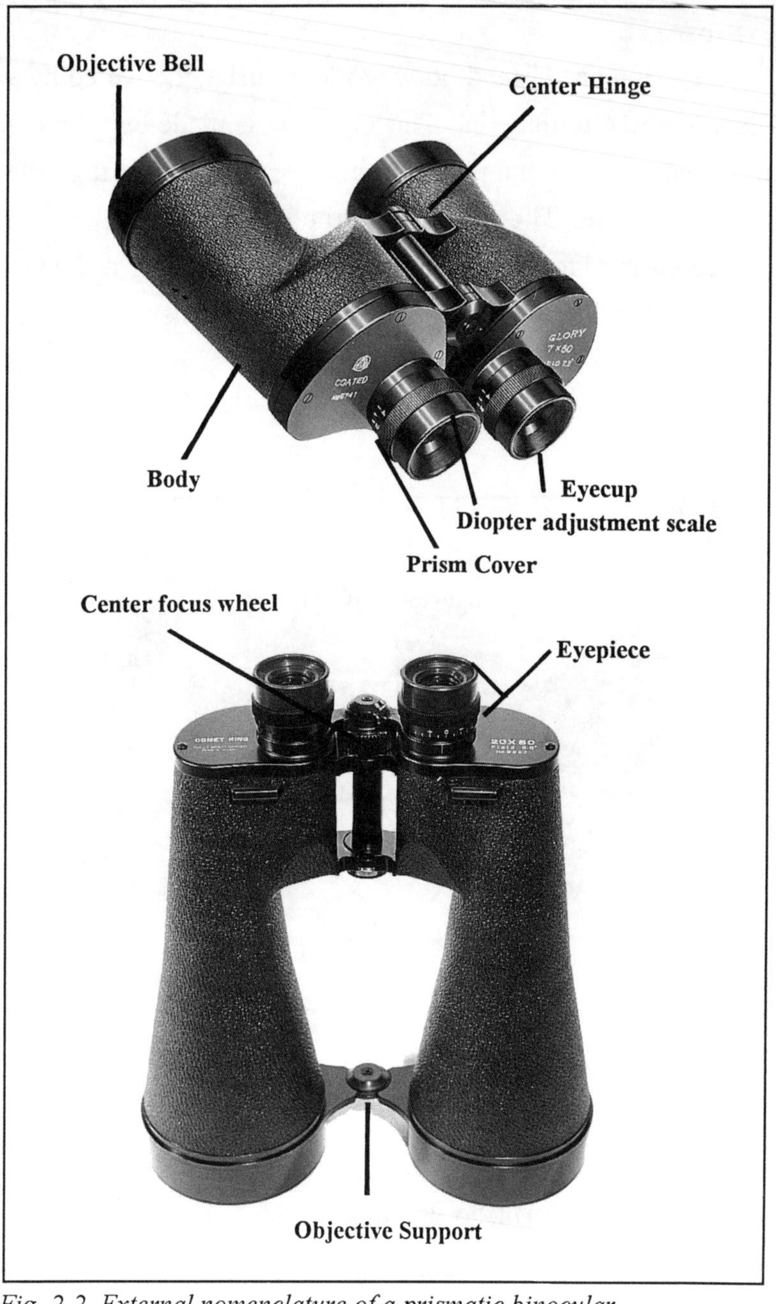

Fig. 2-2 External nomenclature of a prismatic binocular.

The metal barrels and body work are usually covered by a vinyl leatherette material. This covering is made by taking a thin piece of vinyl and placing it in a press with a textured die under pressure. The covering is the carefully trimmed and cemented in place.It has become popular to apply a thicker "armored" rubber coating to give the binocular a pseudo-military appearance. Be warned that just because a binocular has the armored coating *does not* mean it is waterproof.

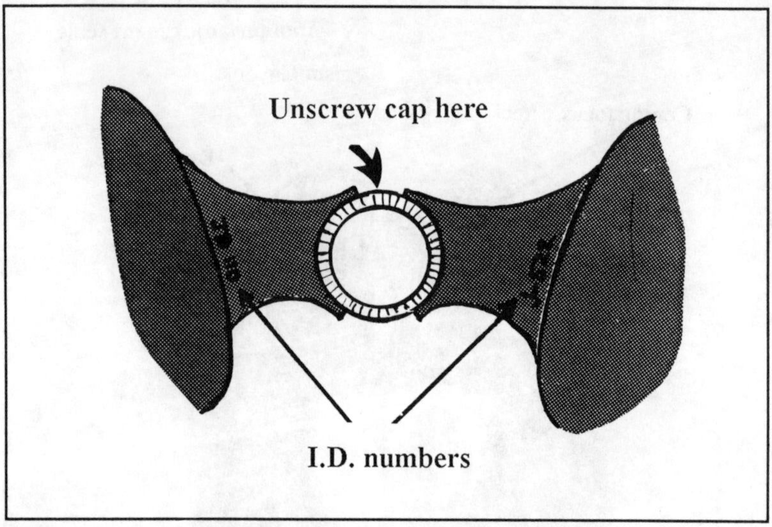

Fig. 2-3 Removable plastic cap covering tripod socket. Also shown are maker's recognition numbers.

On the objective end of the hinge assembly is a small plastic cap that covers a 1/4-20 threaded screw hole. This is where most tripod adapters attach. (Fig. 2-3)

On better quality Japanese models there will be two small sets of numbers stamped into the bridge. The **JL** and **JB** numbers tell us who made the metal parts and who did the final assembly. Typically, binocular assemblers buy their lenses and metal parts from a number of sources. The assembler does the final collimation and assembly and places the exporter's brand name on. It is very rare that you will ever find a maker that does everything "in house."

Finally we come to the ***objective*** end, that is the end of the binocular *closest to the object you are viewing.* Here we find the main light- gathering lenses. These lenses, along with their cells and threaded rings, control the ***collimation*** of the instrument. Collimation means that when we look through a binocular the view is perfectly comfortable with no strain or "uneasy" feeling. Collimation is so important that we have to devote an entire chapter to the subject.

The ***objective*** appears to be a single glass element, but in reality it is two elements cemented together. The objectives are protected by a metal or rubber end cap.

When we move into the chapters on collimation and repair, we will learn how all of these parts can become worn or damaged and how the instrument can be taken apart for repair.

CHAPTER

3

Choosing a Binocular

More misinformation and just plain nonsense has been written about how to select binoculars than any other phase of optical theory. How does this all comes about? It may go like this. A nature or science magazine will ask a technical writer they need an article on selecting binoculars for their upcoming issue. The writer starts doing some research by digging out some old articles and then reinforces these with a batch of manufacturer's literature.

So now you have someone who's only experience may have amounted to nothing more than a glance or two through a friend's binocular at a football game. Pretty soon you've got confusing chart after confusing chart listing twenty different models each one singled out for a specific optical requirement. Has this guy giving all this expert "advice" ever had to use binoculars when his life depended on it? Has he ever stood watch on a ship scanning the horizon for hours on end? Has he ever had to scan across the no-man's land of some God forsaken DMZ searching every rock and bush for enemy infiltrators?

I won't waste your time or add to the confusion by filling this section with meaningless tables. What you will get is the best advice that I can draw on based on over thirty-five years of experience in the field. While some binocular guides are nothing more than manufacturer's literature bound into book form, please rest assured that this will not be the case in this manual.

I would like to relate a little story that happened to me recently that sort of brings it all together regarding selecting the right optical instrument for the job at hand. I had a chance to attend a concert here at the local university featuring a world famous musician. At the last minute I tried to find our opera glasses. You know the type that looks like little clam shell that pops open. They're 2 or 3 power and sometimes you wonder if they do any good at all or it's just your imagination that things look better through them. Well, I couldn't find them. As I rushed out the door, I grabbed my regular 8x30 binoculars in despiration.

Setting back about 45 rows from the stage, I realized how spoiled one becomes from the technical advances television has offered us. Here I was, live at the concert in quite expensive seats, but the performers were so far away that I had to strain to catch their expressions. Then using the high quality 8 power glass I "jumped" right up front. Full facial expressions! I could even read the instrument maker's name! Simply by adding these 8 power-binoculars, I had changed the entire evening from mediocrity to first class entertainment. Yes, the correct optical aid can add many new dimensions to our lives.

Where to buy?

The first question we have to answer is where are we going to buy our new binoculars? For most people it comes down to a retail camera store, discount store, or by mail order. Don't expect a lot of help from the person behind the counter. If it's a camera store, he may know a lot about photographic equipment, but binoculars are usually a sideline . If you are going to make a wise decision the only thing you can do is gather as much information from outside sources as you can. Don't rule out sales brochures as a possible source of information. While you have to take everything with a grain of salt , you can learn a few basic points on construction and terminology.

Mail order presents a few more problems in that the order taker on the end of the phone is not going to be of much help. If you buy by mail, you are going to have to do your homework ahead of time. Make sure they also have a liberal return policy so you can give the item a careful examination. A 30-day trial period for a mail order purchase is not an unreasonable time period for the customer to try out the glass under a variety of conditions

Price and quality

It should go without saying that when investing in an optical instrument be it telescope, camera, or binocular, buy the best unit you can afford. There are no bargains in optics. If you want to pay $35 or $40 for a pair, chances are you're going to get something made in Taiwan, Hong Kong, or some third world country with most of the parts made of plastic, inferior lenses that are not collimated, and prisms made of out of something akin to window glass.

Who makes the best binoculars? Some would argue Germany while others side with Japan. Without getting into brand names, the Germans excel at the roof prism type while the Japanese have the upper hand at producing the traditional porro prism type.

Whatever type you decide on, always insist on a model with BAK-4 prisms and full multi-coating. We find that manufacturers tend to put these features on their best quality models.

The selection process

First, what is the intended purpose of your new binocular and your areas of interest? If they think about it, most people can narrow the field very quickly. Yes, there are some special fields such as boating and marine use, specialized bird watching, and astronomy, but most people want an all around general purpose glass for vacationing and spectator sports. ***Portability and ease of use are of paramount importance.***

Power or magnification

Let's take a good look at power. What do we mean when we talk about the ***power*** or ***magnification***? If we look at an object at say 100yds. away, our reference value is the unaided eye. Now let's put a 7 power binocular in front of the unaided eye. That object out there at 100 yds. now looks like it is only 14 yds. away. (100 yds. divided by 7X = 14yds.) In reality, there aren't really that many choices. 6X, 7X and 8X cover the field very nicely because these are the only magnifications you can hold in your hand without the image dancing all over the place. When you move above 8X, we get into what I call the "jiggle factor" and they must be supported with a tripod. If you realize this fact, there are some special binoculars for long surveillance and astronomy with powers in the 11X to 25X range. Over the years, I have come to realize the binoculars I use the most are usually the lightest and most compact of all those I own. ***Magnification is a minor consideration.***

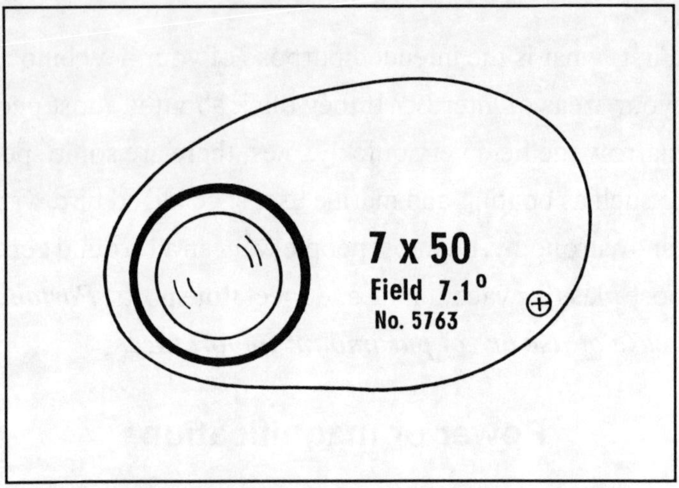

Fig.3-1 The binocular's power is usually indicated on the right prisim cover followed by the size of the objective lens in millimeters.

How big of an objective do you need? The "Age" Factor

How about the size of the objectives? We have all been to football games and seen those macho guys with huge 7x50 binoculars hanging around their neck! I also noticed they are so big and heavy that they are never used very much. Let's divert from every other book ever written about binocular selection! ***The single most important factor in binocular***

selection is age of the intended user! O.K., you say, how does age have anything at all to do with binocular use?

Take a pair of binoculars and hold them up at arm's length pointing out your window at a bright daylight scene. Notice the little round disc in each eyepiece. This is the ***exit pupil*** or Ramsden disc of that binocular (Fig. 3-2). If you care to experiment a little further and run a little optical bench test, you can take a thin sheet of white paper and starting back about 6 inches, slowly move it toward the eyepiece.

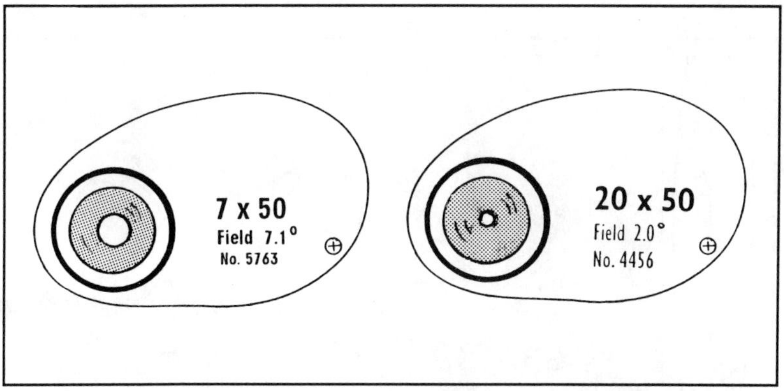

Fig. 3-2 When held at arm's length, the exit pupil of a binocular is clearly visible. The higher the "power," the smaller the exit pupil.

As you move the paper forward, there will be a point where an actual small image of the scene outside the window is projected on the sheet. When this image is perfectly round and sharp, you've reached the ***eye point*** of that instrument.

The eye point shows exactly where your eye is at the

optimum position to get the full benefit of the instrument's field of view and light-gathering capability.

The binocular's ***exit pupil*** is a function of objective lens size and power. If the power remains constant and we increase the size of the objective, the exit pupil gets larger. Try and think of the objective as sort of a "light funnel." In Figure 3-3, the large side of the funnel (A) is pointed toward the object you are looking at, the small end of the funnel "drains" the light into your eye. (B) is the binoculars exit pupil. (C) is the ***entrance pupil*** of the eye.

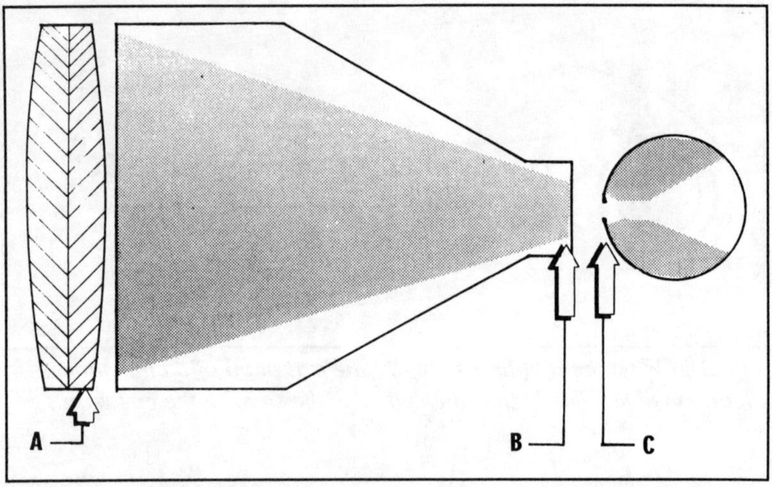

Fig. 3-3 Light from the objective (A) is "funneled" through the binocular's eyepiece (B) to your eye (C).

As shown, the human eye also has its own *entrance pupil.* When you look at your eye in a mirror, the little black hole in the center of your iris is your exit pupil. Remember, it's constantly changing with your environment's light conditions. In bright sun it's small, at night it gets larger. This entrance pupil is also effected by age.

Sometime during the 1940's the Navy ran extensive tests on just how much light the human eye can effectively use. The tested some 17 and 18 year old seamen against some old salts in their 50's. What they found is at night the iris of the human eye opens to it's maximum size. Thousands of years ago, humans were very effective night hunters. As the eye ages, this ability to open and let in more light declines. At 18 years of age, this opening in the eye (pupil size) can reach 7 or 8mm. Add 10 years and it drops down to 6 mm. At age 40 it's down to about 5mm. After 55, 4mm is not unusual. Let's say a person 50 years old picks up a 7x50 binocular. Divide 50 (size of the objective) by 7 (the magnification) This gives an exit pupil a little over 7mm. Since the 50 year old has a 4 or 5mm maximum entrance pupil size, 2 or 3 mm are wasted. The image would have been just as bright for him if he had used a 7x35 (35 mm divided by 7 = 5).

Here are some ***exit pupils for common binoculars***. 6x30 =**5**, 7x35=**5**, 8x30=**3.9**, 7x50=**7**, 10x50=**5**, 8x40=**5.**

Why age is a factor

Figure 3-4 shows the relationship of the *night adapted pupil size to age* along with some suggested useful binocular

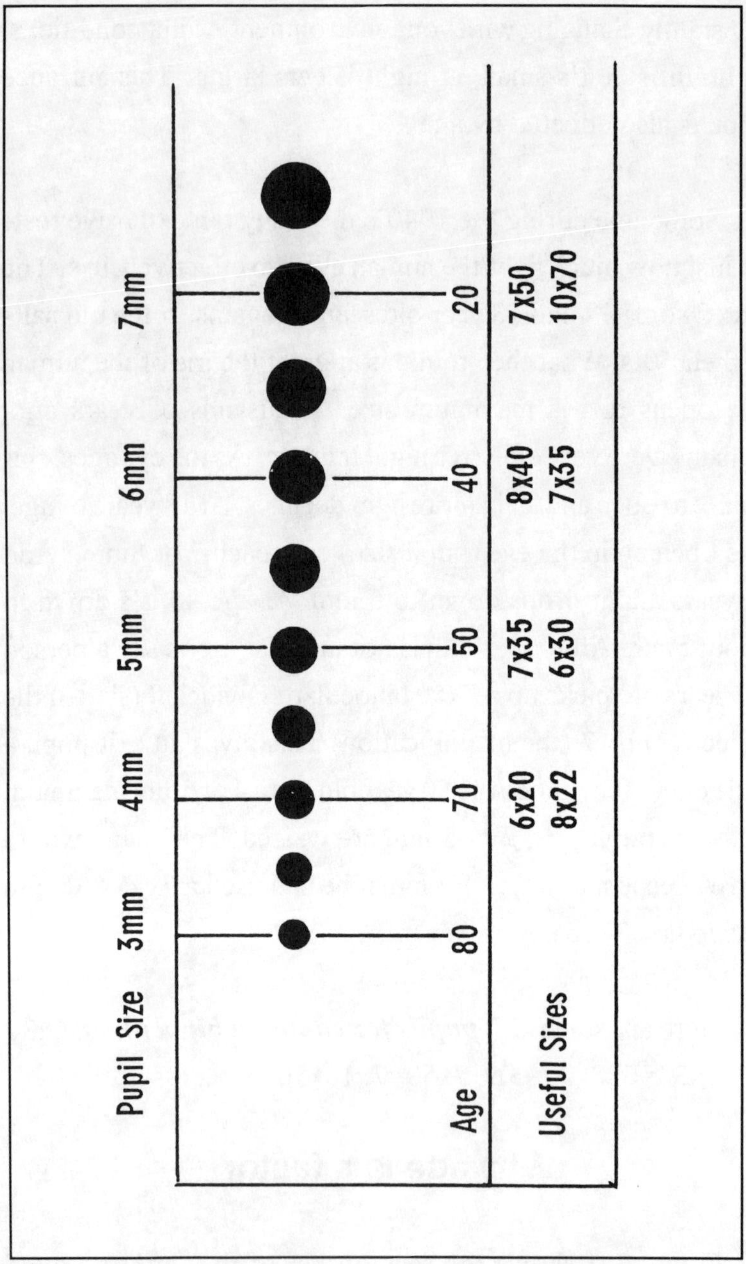

Fig. 3-4 As age increases, the useful size of the pupil decreases.

sizes. Of course, common sense and intended use of the glass should be your guide. While a 20 year old could easily get maximum use out of a huge 10x70mm, if he were going mountain climbing, a lighter weight 6x30mm would be a better choice.

One point to remember, the chart shows the *night adapted* size of the eye's exit pupil. These sizes were measured after a set period in a totally dark environment. In the daylight, the normal size of the exit pupil may go down to about 2 or 3 mm. Therefore, on a sunny day , the image brightness will be about the same if you have a 6x30 next to a 7x50. Your eye just can't use all that light coming at it! ***As a general guide for daytime uses of a binocular, in bright sunny weather, you don't need an exit pupil larger than 2.5 to 3.5mm. In cloudy overcast, dawn and dusk situations, 4 to 6mm.***

Occasionally you'll see the terms Relative Brightness (RB), Relative Light Efficiency (RLE), and Twilight Factor. While I have included the mathematical formulae for these terms elsewhere in this book for the purist, they have little or no practical significance in choosing or using a binocular. Forget them!

If you are the sole user of a pair of binoculars, let your age and the desired magnification guide you. If you are sharing the glass with younger members of the family, it's O.K. to go a little larger on the objective size, but remember, every millimeter of extra glass in the objective means more weight! Glass is heavy.

Special needs of the bird watcher and mariner

Because bird watchers may do a lot of viewing early in the morning when light conditions are poor, most knowledgeable users like to go a little larger on objective size than would normally be the case. A 40 or 42mm objective is a happy compromise in that it lets the eye see more deep shadow detail. A 42mm will let you observe about 15 minutes longer into the dusk and about 15 inches further into deep shadow compared to say a 25mm objective.

The larger 7 mm exit pupil of a 7x50 binocular makes it a little easier to position your eye when the boat deck is pitching and moving. That is why over the years the 7x50 has always been regarded as the premier marine binocular.

Field of view

The ***field of view*** or what you "actually see" (real field) through your binocular is always been touted as a factor in the selection process. Ten or 15 years ago the industry was taken with "wide-angle fever" producing heavy, bulky, expensive instruments with huge 11 or 12-degree fields. If you tried to follow a target or made some panoramic sweeps with them, someone better pass you the motion sickness pills! We are happy to say that sanity has returned and the availability of these models are now limited.

Field of view is conventionally expressed in relation to what we see at a distance of 1000yds. The 1000yd. figure was probably chosen due to the close relationship with the military use of this equipment during WWI in artillery range finding. Today, modern glasses fall in the degree 6 to 8 degree range. You can estimate the actual field in feet at 1000yd. by multiplying the ***angular field by 53*** (Fig. 3-5). (6 degrees X 53 = about 316 ft.)

A point to remember! The field of view is not determined by the size of the objective lens! It's a combination of the *focal length of the objective lens,* the eyepiece design, and the size of the prisms.

All we need to know about field of view is that above normal limits of 6 degrees to 8 degrees, it's of limited value. Yes, there are certain applications, like aircraft spotting, where you want to see a huge slice of the sky. However, as field of view increases, the binocular becomes heavier, harder to hold due to its bulk, and more expensive. You also need to crowd your eye very close to the back lens to see the entire field (reduced eye relief).

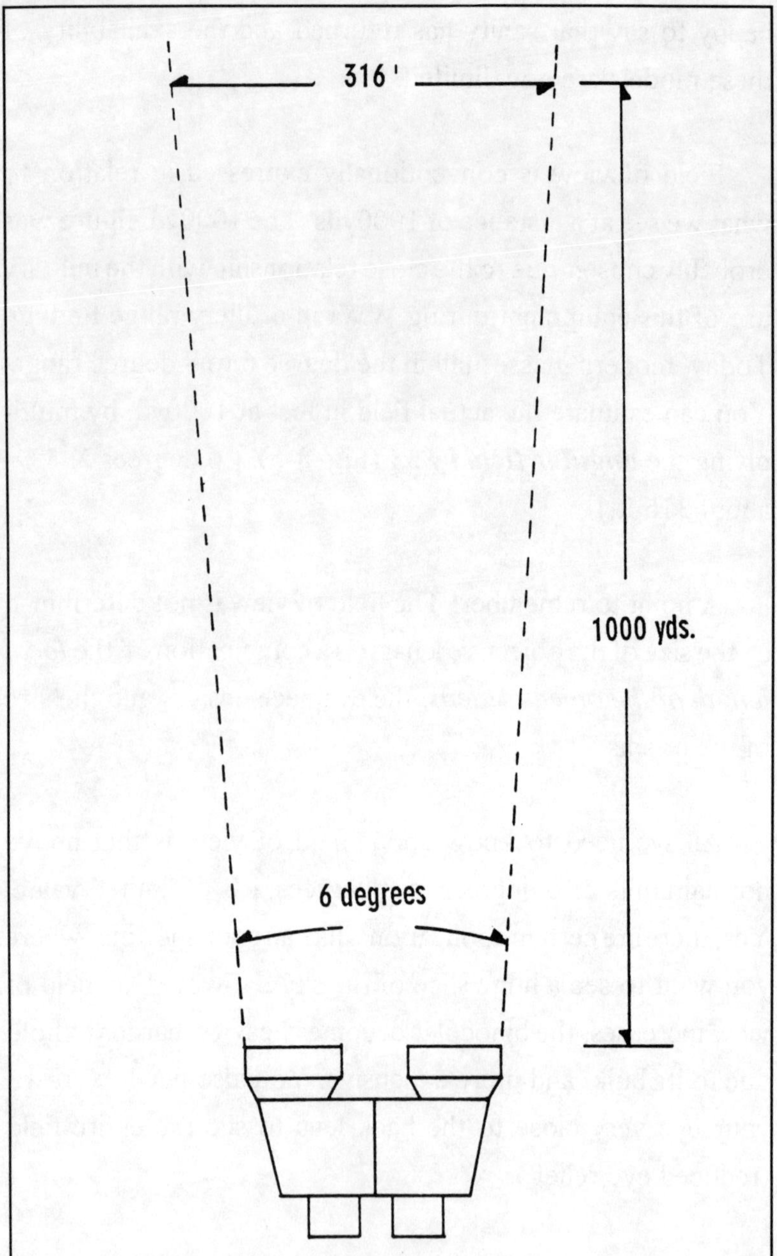

Fig. 3-5 The angular and linear fields of view.

A quick in-store collimation test

The entire complex subject of binocular collimation is cover elsewhere in this text, however when you're going into a retail store to make a purchase, you are on your own when it comes to collimation. The chances are good that the salesperson will not even know what the term stands for.

I will say that among the better brands out there, you will not find many units that are far enough out of kilter to cause you any severe problems. Eyestrain in binoculars is caused by misalignment of the two optical barrel assemblies. If this problem is severe, you can feel it right away! You may or may not see a double image, but in any case you will feel your eye muscles "fight" the binocular.

Take the binocular you're interested in testing out where you can see an object at least 50 to 100 yards away. If it's a center focus unit, be sure you also focus the right eyepiece.

1) Look at your target object. Take the binocular down and let your eyes relax for about half a minute.

2) Now look at the same subject again but cover the *left* objective with your hand.

3) Finally pull your hand away from the left objective.

4) If the object is at first out of line, but then quickly converges again, your eyes are adjusting for the inherent error in collimation and the binocular should be rejected.

Test charts

How good is the image? How "sharp", how fine detail can we see? In other words, how is the *resolution*?

The U. S. Navy tells us that to meet military standards, a binocular should resolve an angle of 4 seconds of arc. Included in this book several test charts modified for binocular testing. They consist of a stepped series of bar patterns arranged from large to extremely small. Cut one of these charts out of the book and mount it on a piece of white cardboard.

Mount the chart outside in bright daylight (not direct sunlight) at a distance of 100 feet. Study the chart with the binocular either tripod mounted or resting on a firm support. While this test can't be done in the retail store, as I mentioned before, you should purchase from someone who gives return privileges. You're looking for the smallest set of bars that are clear and distinct with noticeable separation between the elements. Remember the rule about focusing, from the *out* position to the *in* position.

See how far down the bar scale you can you can resolve the pattern. A good quality binocular should give us detail in B-2, bar 2 or 3. Not quite 4 seconds of arc, but good enough. It's just one element in our entire evaluation procedure along with collimation and general overall fit and finish.

Is there any other way we can check image quality? Not

easily without a little time and effort. You'll find almost 95% of all binoculars out there give a good quality image. Of course some are much better than others, but you really need some kind of test chart where you can sit down and compare image quality side by side. This is just not going to happen in a retail situation! Really, *the only true test of binocular quality is to examine a star image in the night sky.*

A more critical test

If you're ever going to become really proficient at evaluating binocular performance, you need to become familiar with the "star test". This test originated in the field of astronomical telescope testing and evaluation and can easily be modified to testing binocular optical systems.

The absolutely best test of any optical system is to view the tiny, perfect image of a star in the night sky.

A few years ago I was called upon to evaluate some expensive 100mm binoculars. The manufacturer's literature touted these units as being the ultimate in fine quality and they were aimed directly at the critical field of astronomical observation. Esthetically the instrument was stunning. The fit and finish were worthy of a fine camera. Mechanically everything was oil smooth.

Daytime observation looked good. Good sharp images.

Yet somewhere I had this uneasy feeling that all was not perfect. At dusk we mounted the units on a heavy tripod and waited. The moon rose slowly above the horizon. It was best to wait a little longer until we had some bright stars above us for a through evaluation.

Finally we swung this giant instrument on the crescent moon. The image of the moon had a beautiful halo of blue and yellow, an absolutely dead giveaway that the units had severe chromatic aberration or "color." Perhaps all was not lost , and I searched for a field of stars. Holding the star image in the center of the field it looked like a tiny pyramid slightly out of focus! No amount of fine focus could produce anything resembling a pinpoint of light. It appeared that there was an extremely poor quality cut on one of the internal roof prisms. Something like this can be excused in a $20 or $ 30 instrument, but not at this price! Absolutely the poorest image of a star I had ever seen! Nearly $2000 for an instrument that couldn't pass the most elementary optics test!

A good lesson! Don't let high price and and a nice external appearance be your only guide to product quality.

One problem you always run into with the star test is a stretch of bad weather where you can't test them on the night sky. We can get around this by making our own home built artificial star tester using a little .001 or .002 brass shim stock and carefully piercing it with a sharp needle. Make a circular chimney out of heavy foil, glue the pinhole on the chimney, and set it over a low power frosted bulb. (Fig. 3-6)

Once you have your artificial "star," mount the binocular on a tripod and set the star about 40 to 50 ft. away.

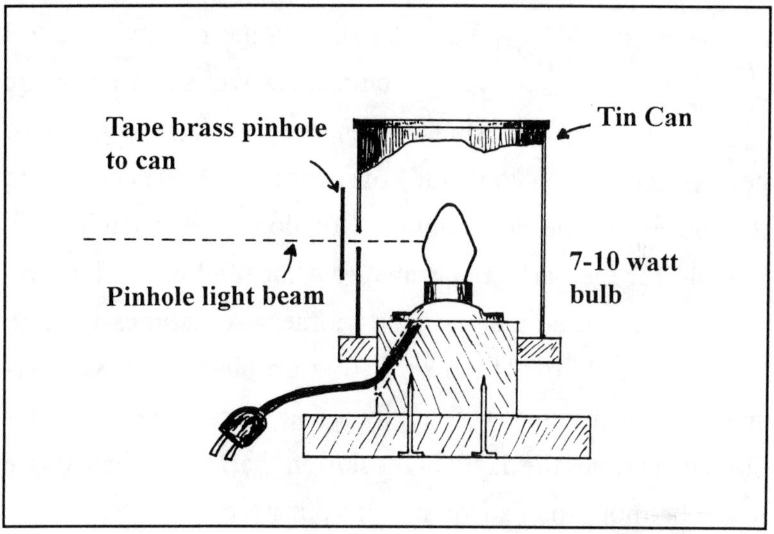

Fig. 3-6 An artifical star for testing can be made out of somthing as simple as a tin can and "Christmas tree" bulb.

To star test, remember to always place the star image in the center of the field of view. Very few binoculars, regardless of price or quality, hold up when the image is placed near the edge of the field. Correctly focus the image using the procedure outlined elsewhere in this book. Figure 3-7 shows an ideal or perfect star image. There is a central disk containing most of the light energy with two or three concentric rings containing the rest of the light energy. Please remember, no instrument including a $3000 astronomical telescope will

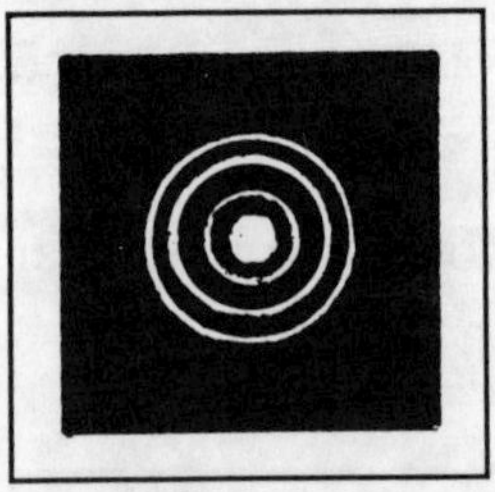

Fig. 3-7 This illustrates a "perfect star" image.

instrument including a $3000 astronomical telescope will show an "ideal" image like this! Figure 3-8 A shows how a pair of poor prismatic binoculars will appear. Figure 3-8 B shows a binocular of average quality. A good pair of binoculars will show an image similar to Figure 3-8 C. If you wish to get more specific in your study of star images, you need a good text book on telescope optics. But don't get carried away looking for the perfect star image. Almost all binoculars will deviate from the "ideal." Astronomers sometimes have to spend upwards of $4000 or $5000 on binoculars with air-spaced objectives to get really good star images. If your binocular passes the daylight resolution chart tests, chances are it will be more than adequate for your needs.

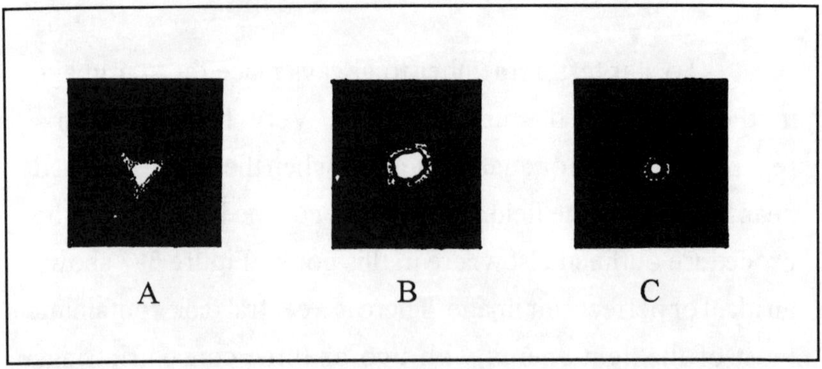

Fig 3-8 A, B, and C The star image is a good gauge of binocular quality. Poor lenses, prisims and collimation all reveal themselves.

As you study some of the product literature on binoculars and look over what's available in the store, you're going to notice two distinctive body styles are available. One type is generally known as the "American" or B&L style and the other the "European" or "Z" type. These terms have absolutely nothing to do with where the unit was manufactured, but refer to the body style design.

The "American" style means the body is cast of one piece from the prism shelf right down to the objective cell. Most military glasses have this one-piece construction because it is very strong and can take rough handling in the field (Fig. 3-9).

Fig. 3-9 The smooth flowing design of the one-piece "American" design.

Fig. 3-10 The "Z" type or "European" style is more damage prone.

The "European" style has the prism shelf cast as one piece and another piece holding the objective lens. These two pieces are then screwed together. (Fig. 3-10) The problem with this type is that a blow to the objective housing tends to knock it sideways a few millimeters and ruin the collimation. The objective bell housing strips the threads in the other portion of the casting. When a "Z" comes into our shop for repair, the first thing we look at is threaded connection. If you give your binoculars hard service choose the "American" style. If you are very careful with your equipment and always keep it in its case, then you can get by with the "Z" model.

In talking with people about choosing binoculars, we always get a few questions about zoom binoculars. They feel that since they have a zoom lens on their 35mm camera and video camera, it would follow that the idea would work well on binoculars.

Zoom binoculars are to be avoided. They are heavy, they are complex, and nearly impossible to repair. Personally I have never seen a zoom that is in collimation or will hold collimation as the power changes.

Conclusion

To tie this all together, purchase a medium to light weight glass, preferably American style, with a normal field of view. If you are going to hand-hold the units, don't go above 8X. If you are the sole user of the instrument, choose an individual eyefocus model. If you are over 40 years of age, don't get carried away buying large exit pupil models, and with the limitations on hand-holding magnification,this means about 8x42 should be your limit. If you plan to hold the binocular on a tripod, you may go to the larger objective.

Conclusion

CHAPTER 4

Learning How to Use a Binocular

Over the years I've had a chance to talk with literally hundreds of binocular users and I am still surprised at how many still fail to grasp the basic focusing process! Now we're talking about people who, through the years have invested in three or four binoculars representing hundreds and hundreds of dollars in equipment.

It may be helpful to actually have your hands on a pair of binoculars while going through the focusing process.

The first thing we want to train ourselves to do, to the point that it should become automatic every time we pick up an instrument , is to *always use the binocular neck strap*. The greatest danger to your glass is to have it bounce off a cement driveway! Camera and binoculars have almost a magnetic attraction to hard surfaces! Train your other family members to use the neck strap if they are using your equipment.

Setting the interpupillary distance

I can't help but be amused when I go to the movies and see some character on the screen using a pair of binoculars to scope out an object. The camera view always shows two circles merged at the center similar to figure 4-1. This is exactly what we don't want to see. I guess we have to allow the movie makers a little artistic license because they assume no one knows what a view through a binocular really looks like.

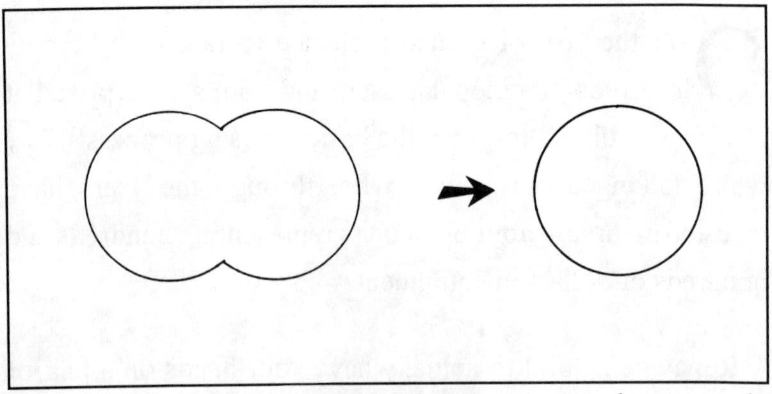

Fig. 4-1 The image from both barrels merge into one by moving the central hinge.

In reality, when we set the interpupillary distance (more simply put, the distance between the centers of each eye) we want the view to appear as if we're looking down a single tube. Grasp the instrument with both hands and gently move the hinge until the images merge. Just as a point of information, the average distance between human eyes is about 64mm. On the top of the bridge between the eyepieces you will note a small scale which usually has two numbers (60 and 70) and a few small divisions. This is the interpupillary distance scale and you can tell just about how many millimeters you are between the eyes.

The so-called experts writing articles and binocular guide books always make a big point of telling you to remember your interpupillary distance so you can go right back to it the next time you use your glass. Are you kidding? Today we're expected to remember our social security number, several bank account numbers, a federal I.D. number and our driver's license number just to name a few! So on top of all this I have to remember my interpupillary distance setting is 68 1/2 mm?

Professional binocular users always permanently mark the correct setting on their instruments. I first saw this at an outpost years ago on the Korean military DMZ. Guys whose lives depended on rapid identification of targets always indexed their binoculars. Some had gone as far as to drill and run a small pin right through the hinge so no movement was possible.

The best way is to simply take your pocket knife and make a small reference mark (Fig. 4-2). If the idea of scratching your glass offends you, use dots of white ink.

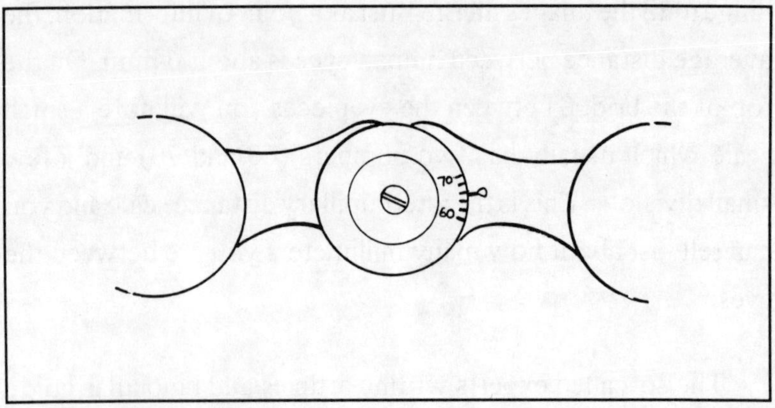

Fig. 4-2 The interpupillary scale shows you the proper spacing for your eyes.

Now that we have the correct spacing, we can tell you that there are two basic focusing designs. The I.F. or individual eye focus method and the C.F. or center focus design. While the center-focus design is very common and is found on most binoculars sold today, from a structural standpoint it is also probably the least satisfactory. It's also impossible to water-proof a center focus design,

Focusing an "I.F." binocular

Let's start with the I.F. model. The focusing method is very simple. Most military optics have the I.F. design. The eyepiece holder has an external thread and the eyepiece has an internal thread. (Fig. 4-3) Movement of the eyepiece or focusing is accomplished by turning either the right or left eyepiece with your fingertips. You'll find "diopter" markings running from (+) to (-). These act as reference points after you have reached perfect focus so you don't have to go through the entire process every time.

Fig. 4-3 Focusing an individual eye focus model.

O.K., here is the important part! Pick an object at about the distance you want to use the glass, usually infinity, and focus from the plus to the minus position. That is, from the extreme "out" position inward. Back the threaded eyepiece way out and then focus forward or "in." The image will be out of focus at first and as you move the eyepiece inward it will slowly come into sharp focus. When it's sharp, stop! Don't try to focus by a "rocking" action. I know, that's the way we all learned to focus instruments. You reach the sharp point and then move back and forth to fine tune the focus. For me, learning the correct way to focus an instrument came after reading about some work done by the Navy years ago.

During World War II, naval researchers ran some very comprehensive tests. They wanted to know the best and most precise way to focus not only binoculars but optical gun sights and range-finder telescope . Remember, this was way before sophisticated radar and human eyes were the only way to spot the enemy. At the beginning of the war, it was not uncommon for Japanese ships to avoid detection because they could see our vessels long before we could see them. Their secret? Large 120 and 125mm naval binoculars built with German technical assistance.

The navy's tests clearly showed that when you focus by coming from the minus to the plus, (starting at the "in" position and then moving "out") serious errors resulted. It seems the human eye tends to overcompensate or "accommodate" and the tiny muscles allowing the lens in your eye are under stress. Image sharpness is never as precise and eyestrain and headaches often result.

When I first read about these test results, I had my doubts. Using a very precision military I.F. binocular, I tried focusing on a test chart my "old" way and then the Navy's way. My way resulted in errors as much as 1/2 diopter. That's about .003 to .005" inches of actual eyepiece movement! Getting back to focusing the I.F. model, it is often helpful to close one eye while you try and focus with the other eye. Close the right eye, focus the left eyepiece. Close the left eye, focus the right eyepiece. Then open both eyes for a perfect picture!

Focusing a "center focus" binocular

O.K., so now we know the correct way to focus and a little about the theory involved. Now let's move on the center focus design. In the center focus binocular, focusing is accomplished by a center wheel that moves both eyepieces at the same time. Some models use a "rocker" or ratchet device, but the principle is the same. (Fig. 4-4)

Fig 4-4 Center focus binoculars are focused by using a central focusing wheel.

Select a distant object. Turn the focusing wheel counter-clockwise until the eyepieces move all the way out. You are now in the extreme "plus" position. ***Close the right eye*** and slowly move the focusing wheel clockwise until the view through the left eyepiece becomes sharp. If you go past the critical focus point, stop, back it all the way out again and start over.

Now that the view through the left eyepiece is sharp, ***close your left eye*** and open your right. Don't touch the focusing wheel from this point on! The right eyepiece is focused by turning the right threaded eyepiece. Turn it all the way out, then slowly focus "in" until critical sharp focus is achieved. Open both eyes and you should have a perfect view .

Anyone who uses binoculars knows that if we are watching an auto race or looking at birds, the distance to the subject is constantly changing. Since we have already correctly focused the binocular for each eye, all we have to do is touch the center wheel a little now and then if our subject moves nearer or farther away.

Just to review, try to remember this little phrase: ***to bring an object into focus, focus in.***

Fixed focus and auto-focus Binoculars

What about some of the new "fixed" focus binoculars now appearing on the market? They say you don't need to focus at all. Sounds great. They set their focus at the factory at a distance slightly less than infinity. Using an eyepiece with a little longer eye relief, the eyes are forced to adjust or "accommodate" to subjects at lesser distances. They work O.K. for young eyes, but if you're over 40, forget them! Now don't confuse these with " auto-focus" binoculars coming onto the market. These new style binoculars have an electronic sensor similar to an auto-focus camera. It senses the actual distance and activates a small servo motor that moves the focus mechanism. It appears to be fairly complex and only time will tell how they hold up.

One last word on instrument focus. As we mentioned above in reference to the fixed focus binoculars , you can bring a target into focus by forcing your eyes to accommodate. This also happens to some extent even while you're attempting to focus the eyepiece. The professional observer trains himself to completely relax the eyes before attempting to focus. Personally, before I focus a binocular or telescope I always close both eyes for about two or three seconds. If you can't get a good focus after about 5 seconds, repeat the resting phase. Do not "squint" the eyes or serious errors will result.

CHAPTER 5

General Disassembly Procedures

When a binocular comes into a repair facility the first thing that's done is to give it a general visual examination. You can pinpoint a lot of problems just by visually looking the unit over. First hold the unit by the objective end and with the eyepieces pointed at a light source and check for any internal fog, dirt particles, or loose screws or glass chips.

Make a mental note of any external cosmetic damage. Are the metal name plates dented or nicked? This will give you an idea as to what kind of care the owner gave the unit. If there is a lot of external damage you can almost be sure the collimation is off and maybe even the prisms are out of alignment.

The " Z " joint

Move the center focusing wheel back and forth. The eyepieces should move back and forth in a smooth motion without catching or sticking.

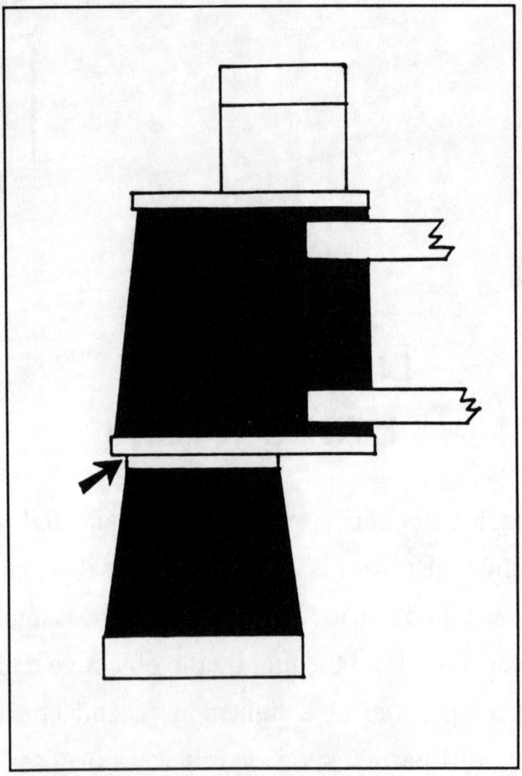

Fig.5-1 The "Z" joint is a potential mechanical problem.

If it's the German style with the objective lens cones screwed into the main body, check the threaded connection carefully. (Fig.5-1) If there is just a few millimeters of space on one side, then the whole objective housing is crooked and must be unscrewed and re-threaded by hand back to its original position. On occasion you may find a tiny screw right through this threaded joint. Remove this screw before you attempt to re-thread the housing. (Fig. 5-2) Be prepared to use some real force to get this joint apart. On large binoculars it's sometimes helpful to have a helper hold the body of the binocular while you grab the bell housing with both hands and give it a real twist.

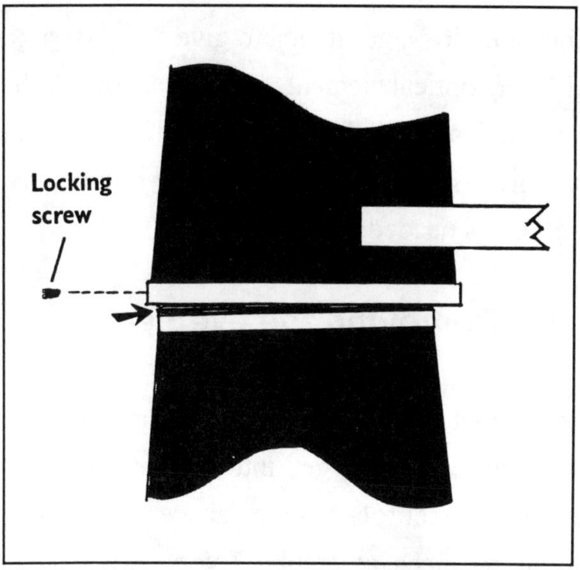

Fig. 5-2 Remove locking screw (if present) before unscrewing objective barrel.

After you have unthreaded this joint, carefully clean out the threads on both pieces with a toothbrush and carefully screw it back together. Re-threading the objective cone back into the body can get tedious. Due to the way binoculars are designed, the female thread in the body has 30% to 40% of the thread cut away. This thread is also very fine so you need to use great care and patience to get the thread started correctly. Run it in three or four threads, and then "eyeball" it to see if it's straight. If it starts cross-threading, back it out and start over again. When you're done, of course the binocular must be recollimated. After repairing several hundred of this “Z” type construction, the best advice to first-time binocular purchasers as to go with the one-piece body style.

At this point it's a good idea to give the glass a good hard shaking! If any optical elements are loose, you can hear them jumping around in their cells. These would include the eyepiece lenses, the prisms , and the objectives. If you find any that are loose, tighten as needed.

Removing the eyepieces

Let's say that an eyepiece lens or prism was a little loose and it was necessary to get into the eyepieces or prism chamber. On a center focus binocular we are going to have to remove the eyepiece assembly. Every amateur repairman makes his first big mistake by attacking this problem from the wrong end! In this case the "wrong end" is the little screw under the interpupillary marking scale between the eyepieces. (Fig.5-3) When you remove this little Philips head screw and the plastic cap, you run into a slotted screw that holds the tension on the hinge. *Don't touch this joint!* This hinge tension was set at the factory and if you mess it up, it's not going to be easy to get it back to the right setting.

Fig. 5-3 Do not attempt disassembly from this end of the instrument.

The correct way to remove the eyepiece assembly on a center focus binocular is through the distal end of the center hinge. That's the end on the objective side of the instrument.. (Fig. 5-4) Most likely this access hole will be covered by a small plastic or threaded aluminum cap. Unscrew the cap. You will find a hole that appears quite deep with no bottom. However, if you shine a small light into it, at the bottom you will

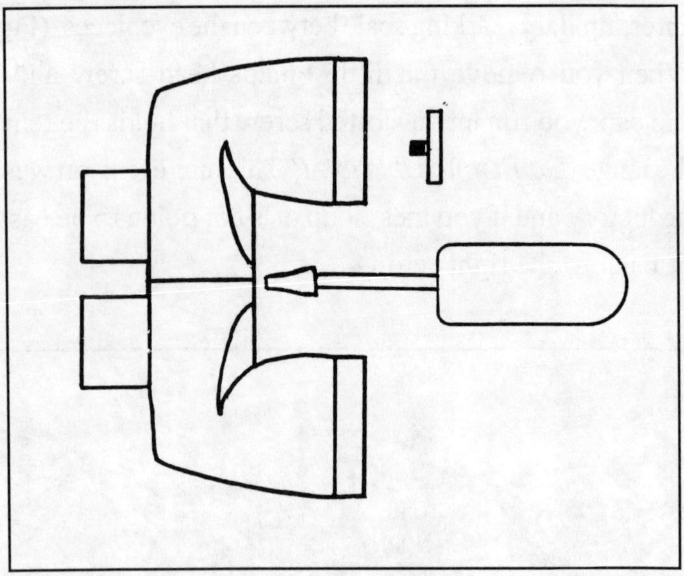

Fig.5-4 Remove plastic cap and insert screwdriver.

find a brass screw with a slotted head (Fig. 5-5). To reach this screw you will need a screwdriver with a long and narrow shank. It may be necessary to take a regular screwdriver and grind it down so that it will slip down into the hole to reach the screw. After you loosen this screw , turn the center focus wheel in a counter-clockwise direction and eventually the whole eyepiece assembly will come loose in your hand. (Fig.5-6)

If there are loose or damaged lenses in the eyepieces they can easily be reached at this time by removing the retaining ring in each eyepiece. ***An important point!*** Binocular eyepieces can be fairly complex. They may contain anywhere from 3 to 6 elements. As each element is removed, set it aside in the order in which it was removed. An experienced repairman will also

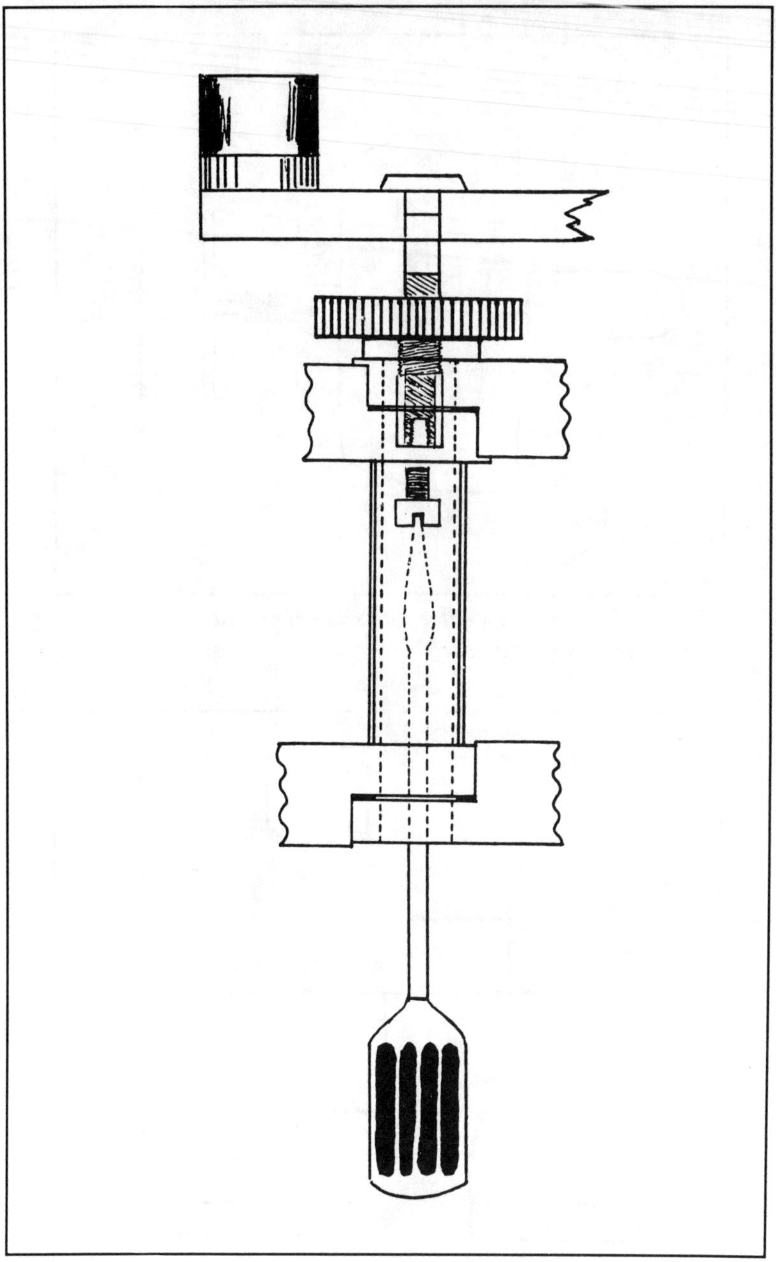

Fig. 5-5 Remove brass screw to free eyepiece assembly.

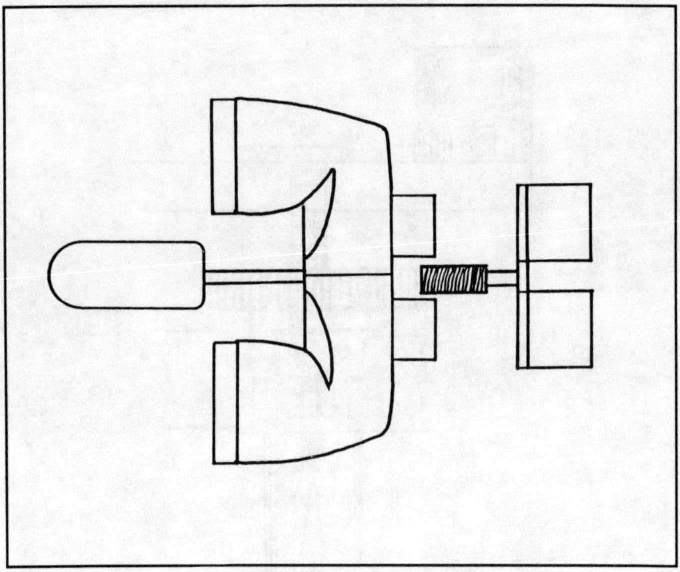

Fig. 5-6 Turning the center focus wheel counter-clockwise will remove eyepiece assembly.

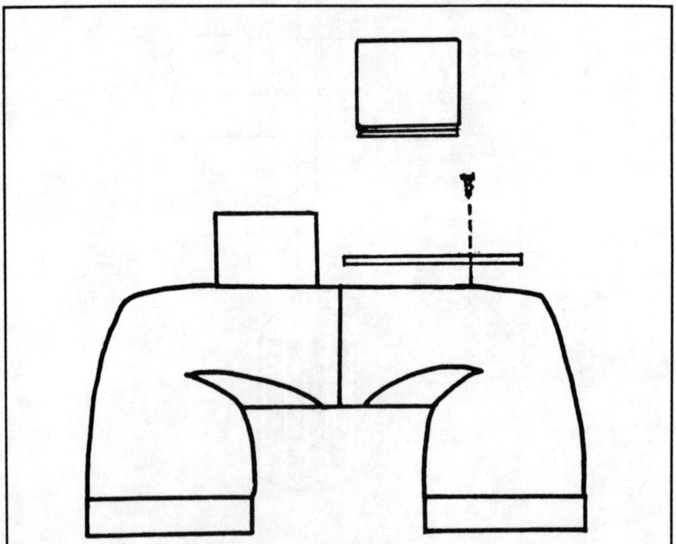

Fig.5-7 Remove the prism cover plate.

make a quick sketch of the lens noting the lens curvature on each side as well as which side goes toward the eye and which side goes toward the objective. There will also be small metal spacers between some of the elements. Carefully note their positions.

Setting aside the eyepieces assembly, we are now faced with opening the prism chamber There are two eyepiece *guide tubes* that must be unscrewed before we can remove the prism cover plates. These have to be unscrewed in a counter-clockwise direction and they can be hard to get hold of and turn. Sometimes a few wraps of "duct" tape gives you a better gripping surface.

After the guide tubes have been removed, the *small Phillips head screw* holding each prism cover plate can be unscrewed. (Fig. 5-7) Removal of the prism cover plate reveals the prism chamber. Removing and adjusting the prism shelf is covered in the chapter on collimation.

To put the instrument back together, we have to reverse the above procedure. The only tricky part is putting the eyepiece assembly back through the center focus wheel. Please refer to figure 5-5. The center focus wheel screws into the hinge assembly. The eyepiece assembly then screws into the center focus wheel. Your basic problem here is that when you turn the center focus wheel all the way clockwise as far as it will go that the two eyepieces must reach their *maximum forward travel position.* If they do not go all the way forward, you may have trouble focusing the instrument. This is pretty much a trial and error procedure. Start by threading the central focus wheel into the hinge assembly about half its thread length. Then turning the center focus wheel in a clockwise direction, thread in the eyepiece assembly. Continue turning the center focus wheel until it bottoms out. If the eyepieces have not moved to their maximum forward position, you have to back the whole mechanism out, turn the center focus wheel a few turns, rethread, and try again. It may be necessary to repeat this procedure 5 or 10 times before you get it right! That is, as the center focus wheel bottoms out, the eyepiece assembly has moved to the maximum forward position.

The central hinge

Just a few important points about the center hinge that we uncovered by removing the small plastic end caps. Over a period of years and with constant use, this hinge may loosen. When you hold the binocular by one side, just the weight of the other half is enough to cause that side to slowly swing down. You can tighten it by applying clockwise pressure with a large flathead screw driver on the screw between the objectives. Get to it by removing the plastic cap with the interpupillary marking on it. On some earlier models you will find a tiny brass locking screw that has to be loosened before you can make this adjustment. (Fig.5-8) *The tension on this hinge is critical!* If it is too tight when you adjust the center focus wheel , the

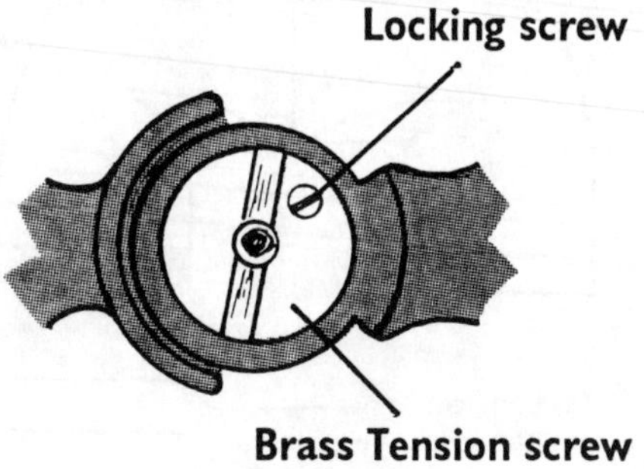

Fig. 5-8 Tension on the center hinge is adjusted by the brass tension screws.

eyepieces will constantly "catch" on the eyepiece guide tubes and throw the entire instrument out of collimation.

You want it just tight enough so that when the instrument is held up on one side the weight of the other side will not cause it to slowly sink down.

Taking out the objectives

Removal of the objectives is quite simple. There is usually a smooth finish ring that must be unscrewed by hand. Some people refer to it as a "dew cap." Upon removal, you will find a slotted retainer ring. (Fig. 5-9) This holds in the objective cell and also locks the lens at the correct collimation setting. Take a spanner wrench and loosen and remove this ring. Then hold the binocular upside down and gently shake it until the lens cell holding the objective falls into your hand. Turn this cell over

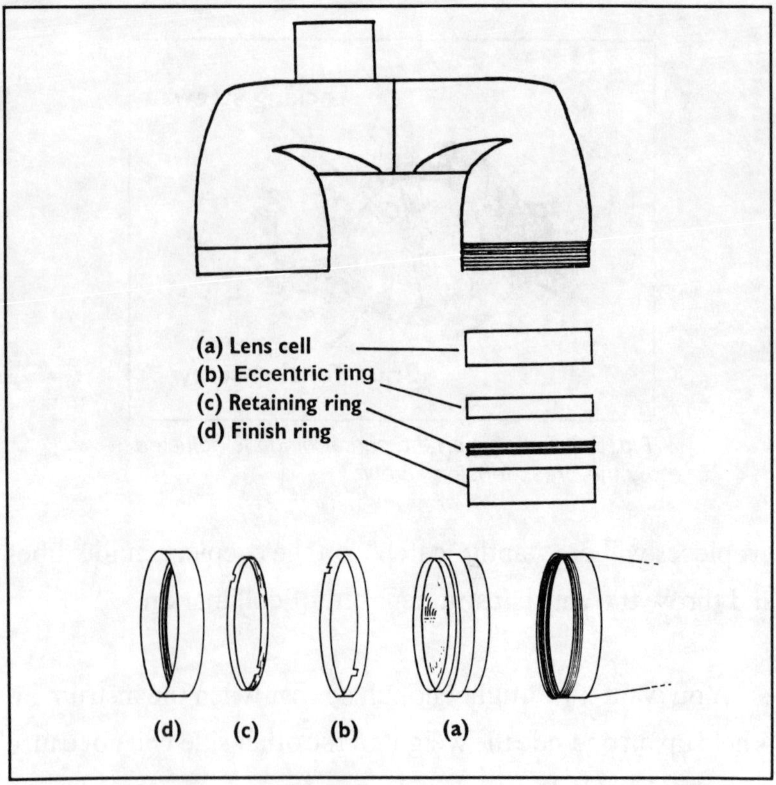

Fig.5-9 Removing the objective cell.

and you will find another retainer holding the glass lens in the metal cell. Be careful that when removing the lens from the cell you make a pencil mark on the edge of the lens with the arrow pointing forward. In general, the side of the lens with the greatest convex (bulge) faces forward.

To review, a binocular has three main assembly groups. The eyepiece assembly with the center-focusing wheel, the prism assembly, and the objective lens with its cell and housing. While each of these sub-groups need to be dealt with separately, we always have to remember that they must work together in concert to produce a perfectly functioning instrument.

CHAPTER

6

Collimation

One area of binocular repair that deserves considerable attention is optical collimation. No matter how good the instrument looks on the outside, no matter how precise and beautifully coated the optical components are, once we put the glass to our eyes everything else pales if the alignment or **collimation** is not precise.

By definition, collimation is the careful alignment of the optical axes of an instrument be it a binocular, telescope, or microscope. At least 90% of binoculars arriving at a repair station are there because of collimation errors!

For a binocular to give satisfactory performance, the image through both sides must be exactly alike. If not, the human eye has the ability to force itself or to accommodate to the misaligned images. The eye muscles sort of go on auto pilot and we don't even know it. They can hold the eye in unnatural positions for long periods with resulting stress headaches.

To get a better idea of what we're talking about and to lay down a little background information, let's start with a straight hollow brass tube lying on V-blocks. (Fig. 6-1A) It does not contain any optics. With our eye centered as best we can while peering down the tube, the tube is slowly rotated. You quickly

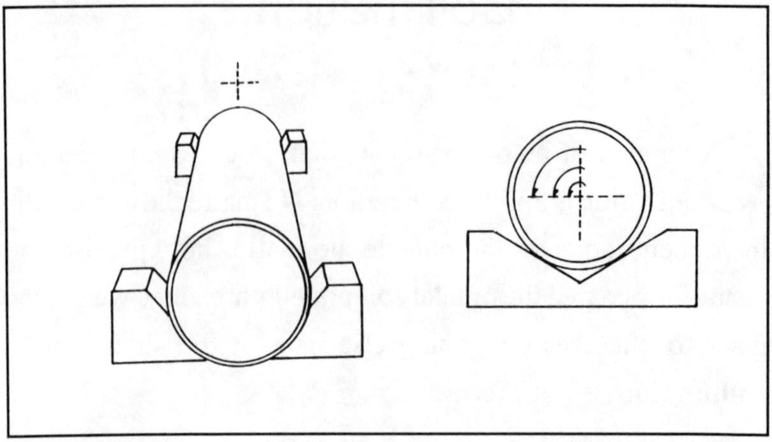

Fig. 6-1A & 6-1B
A simple hollow tube can demonstrate the principal of centered optics.

notice that the *least* amount of rotation is at the *center* of the tube. (Fig 6-1B)

Without being particularly careful, let's glue a set of crosshairs on one end of the tube. Rotate the tube again and you note that if those crosshairs are not precisely in the tube's optical center, they will move in a circle. (Fig.6-2A) If they had adjustments, we could move them around until they were perfectly centered. If we could duplicate this tube exactly and set them side by side, the concept of what a binocular is should start to emerge. (Fig.6-2B)

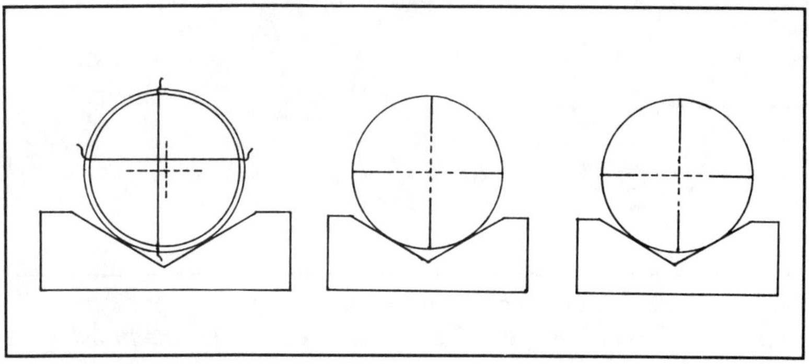

Fig. 6-2A & 6-2B
Adding crosshairs shows the center of the tube's axis.

Now an optical lens also has an optical axis. We can't see it the way we can a crosshair, but its like an imaginary line running straight through its center. (Fig. 6-3). If we added a lens at both ends of our brass tubes mentioned above, we would have a very simple optical telescope that would form an image of a distant object. Again repeating our test procedure of rotating the tubes in the same direction, we may find the image of our distant object seems to "float" or revolve around the tube's center much like our example of misaligned crosshairs. Why? The optical axes of the lenses are not aligned with the mechanical axes of the metal tubes. But if we could move the lenses in the tube in a lateral direction making careful adjustments while viewing the target, finally we could arrive at a point where there would be no movement of the image while the tube spun.

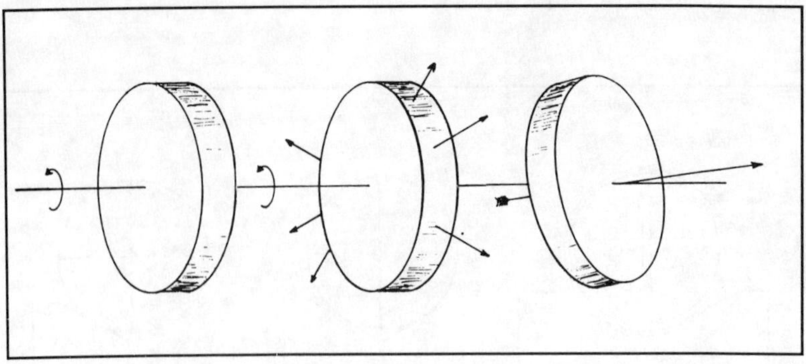

Fig.6-3 The optical center of a glass lens.
Fig. 6-4A Binoculars are collimated by moving the lens in a lateral direction.
Fig 6-4B We do not collimate by tilting the lens.

This is the basic concept behind binocular collimation. An important fact to remember is that a binocular is not collimated by tilting the lenses, but by moving them in a lateral direction (Fig.6-4B)

A binocular's best performance is gained only when the

optical axis of the objective lens passes through the optical center of each *lens* and *prism*.

In the real world we have to make a huge assumption that the metal work of the instrument we are collimating is perfect in respect to correct angle and concentricity of machining. Now this may not be true, but none of us have the time or equipment necessary to go through every metal component, prism shelf, lens cell, and body housing checking for perfect tolerances.

What are some of the alignment problems we can run into when testing or evaluating the performance of a glass? This is

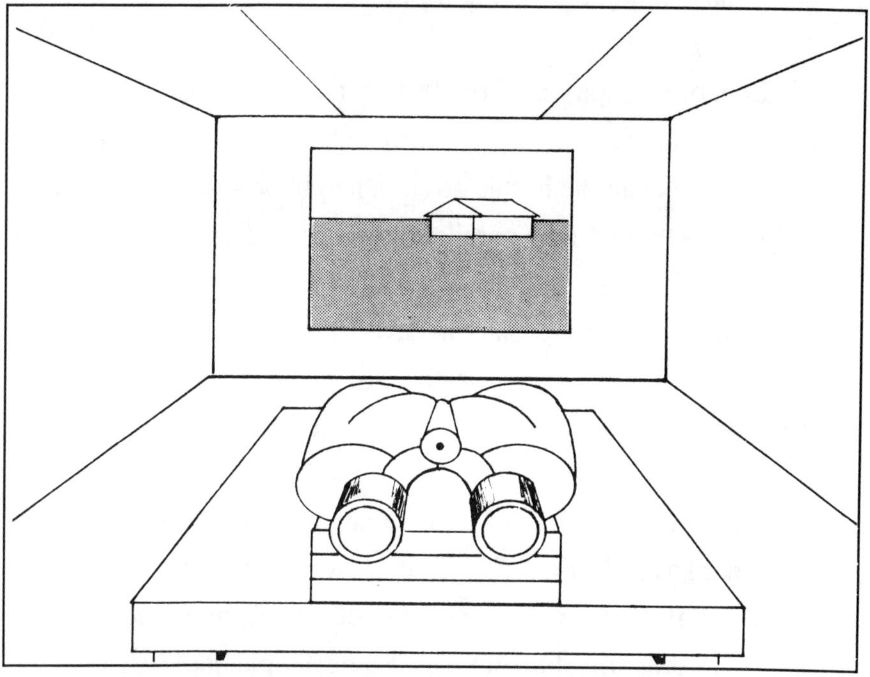

Fig. 6-5 A simple home test for checking collimation.

handy to know even if you are only going out to a store to buy a new pair of binoculars. While it's very difficult to do a careful and complete test with the sales person standing there, sometimes you can spot some of the more serious faults. In a retail store situation, you really need to see out a window at objects 50 to 100 yds away.

At home, it's a little easier. Grab a pair and take them outside. A sturdy picnic table makes a good test bed in that we can rest our elbows on the table for a steady support. Mounting the optics on a camera tripod is also a good idea. (Fig.6-5) illustrates a kind of neat setup using a heavy table and shooting through an open window. The point is that you don't want the binoculars sliding or moving around.

Locate a test target 75 to 100 yds. away. It's best if we can locate a building edge, a roof with a chimney, or a telephone pole. In the example in this book, a neighbor's garage with a strip of black duct tape parallel to the ground was used.

With the test binoculars as level as possible, put the test object in the bottom of the field of view at the 6 o'clock position. Now close your right eye and study the position of the target. Some folks find it uncomfortable to hold their eyes closed for long periods of time so an alternative is to hold a black card over the right objective. Now close your left eye and repeat. With a little practice you can alternately open and close each eye comparing both images. Some people find it easier to just use one eye while shifting their whole head back and forth in front of each eyepiece.

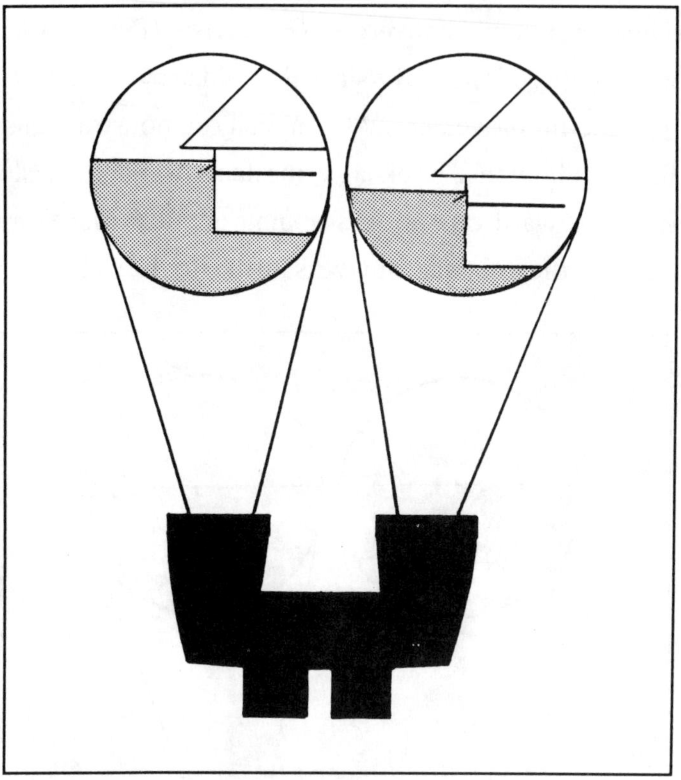

Fig. 6-6 Vertical alignment error

Getting back to the binocular under test, if one image is higher than the other image, then the optical system is "off" in vertical alignment, or is said to have a vertical error(VE). (Fig. 6-6) An error in vertical alignment is one of the more serious problems because the human eyes have little muscular freedom of movement when one image lies above or below another.

Now move your test target over to the 3 o'clock position and repeat the above "blink" technique. If the image in your right eyepiece is slightly to the left of the image on the left side, you have a horizontal convergent (HC) error. (Fig. 6-7) A little error in this direction is O.K. since the human eyes are designed for this "toe-in" movement. We can easily demonstrate this fact by first holding this book at arms length and then slowly moving it toward our nose, stopping about 6 inches away. However, if the right hand image is further to the right, we have

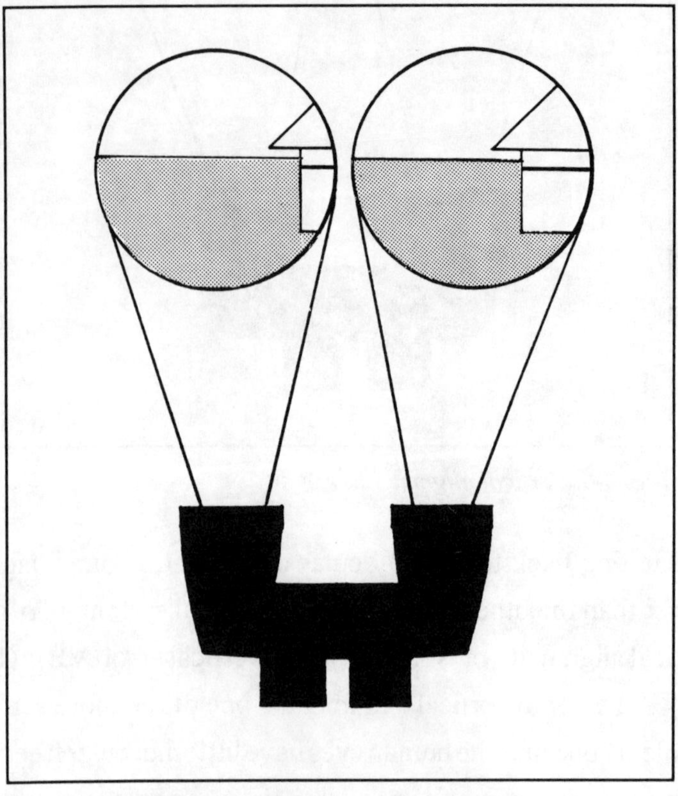

Fig. 6-7 Horizontal convergent error

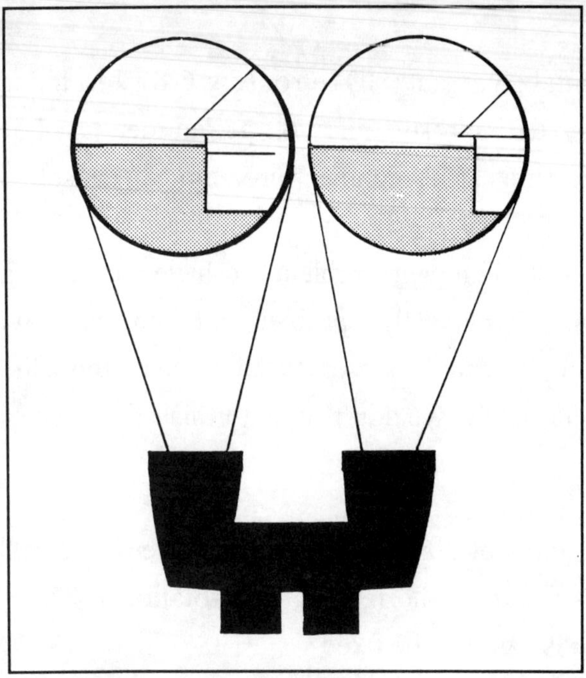

Fig 6-8 Horizontal divergent error

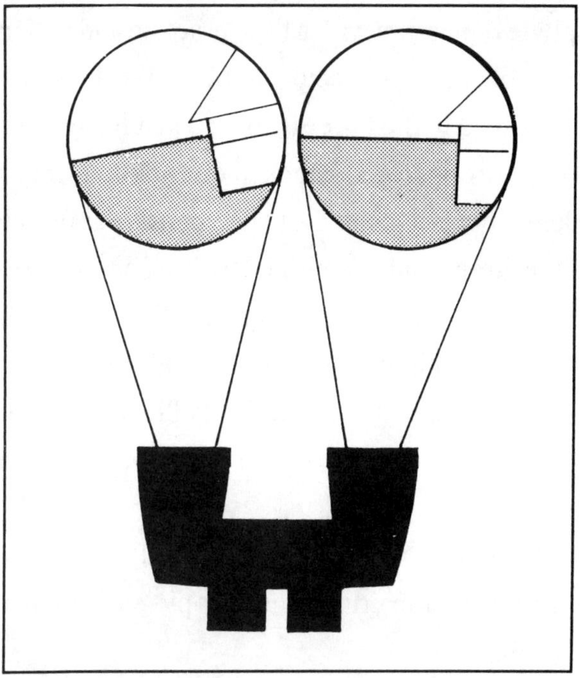

Fig 6-9 Rotational error

a horizontal divergent (HD) error (Fig.6-8). Just try and force your eyes to " spread" and you get an idea just how much discomfort this little alignment problem can cause!

The last alignment problem we have to worry about is rotational. This is really a problem in prism squaring and has nothing to do with the lenses. We'll touch on this a little later. (Fig.6-9) Luckily, we don't run into this very often.

Just to review, binoculars can have three basic collimation problems: vertical, horizontal, and rotational. The next big question is can they be fixed?

The answer is yes, optical alignment is within the grasp of any neophyte repairman with a little practice and patience. The first thing you have to decide is how deeply you want to become involved in the process. If you just have one pair that the kids have dropped and knocked out of kilter, then you will want to keep things as simple and inexpensive as possible. This is called the "free hand " method. By free hand we try to align the glass without the aid of any professional collimating equipment. Basically we just follow the previously outlined process that we used to check for collimation errors while making corrections in the position of the objective lenses.

If you have several binoculars that need work or are willing to try your hand a repairing other people's equipment, you'll need to scrounge around for some common optical items including either building or buying a collimator. Having an optical collimator kind of puts us in the professional class because it frees us to work indoors day or night without worrying about the weather. One essential tool we need before we can do any work is a spanner wrench. Check the chapter on tools about either buying or making one.

Getting started

If you are willing to take the plunge, you need to know that there are several ways to adjust the optical axes of both sides

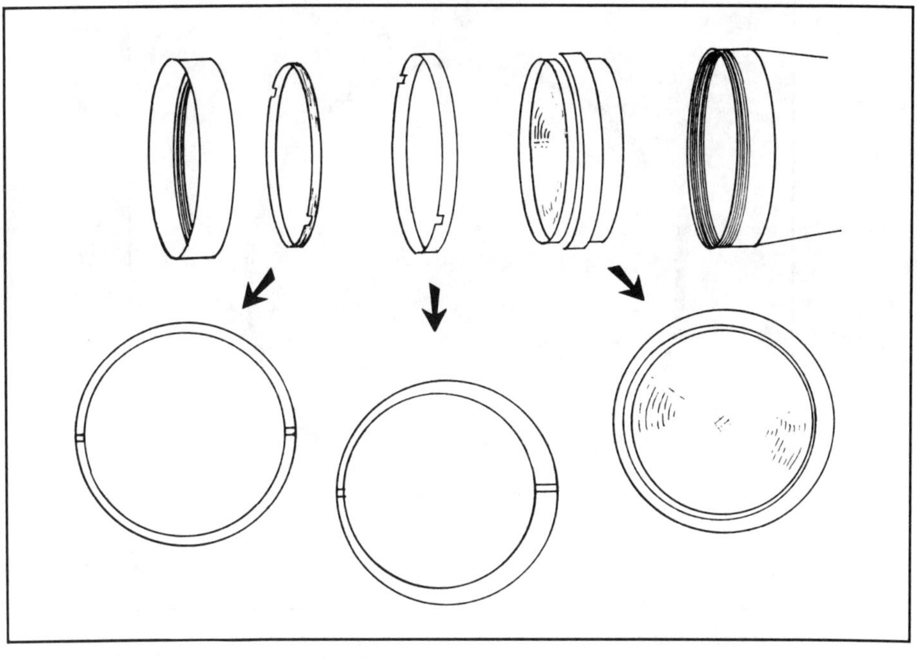

Fig 6-10 The objective lens is mounted in an eccentric retainer ring.

of a pair of binoculars. In some older WW-II models made for the army, we actually had to get inside and adjust the prism shelves. Fortunately modern instruments have done away with prism adjustment in favor of lateral movement or shifting of the objectives. Remember, the objective lens is that lens closest to the object being viewed. The objective lens is always mounted in some type of cell.

There are two types of lens cells in modern instruments. Type E (Fig. 6-10) puts the lens in a set of eccentric rings. By moving one or both of these rings, the lens moves in all

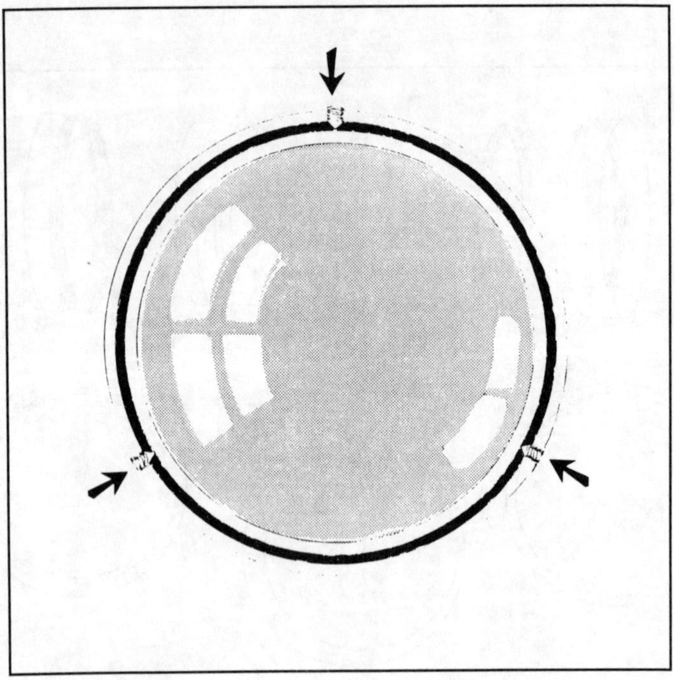

Fig 6-11 The objective lens moves by three pressure screws.

directions. This type of cell is found in perhaps 95% of all binoculars. A second type (A) is found in binoculars using very large objective lenses (70mm or larger). A simple round cell floats on 3 adjusting screws. By loosening or tightening any of the screws, the lens is shifted. (Fig 6-11)

To determine what type of cell your binocular has, you first have to unscrew the dew cap. The cell is fastened tightly into the binocular body with a retaining ring. Before any adjustments can be made, take a spanner wrench and loosen the ring.

Starting with the simplest and also the least accurate "free hand " method, mount your binoculars on a tripod or use a table set-up as shown in figure 6-5. Work outdoors or through an open window. Why an open window? Window glass varies in optical quality from terrible to mediocre at best. An average quality window will distort light rays passing through it in most distressing ways.

Adjusting for Vertical Alignment. After locating our test target at least 75 yards away, and having noted that the right image is a little higher than the left image, we have to make a corrective adjustment. ***To lower a binocular image, we must move the objective lens up. To raise the binocular image, we must move the lens down.*** (Fig. 6-12)

Turn the concentric ring on the right barrel so that the thinnest portion is toward the top. This is moving the lens *up*

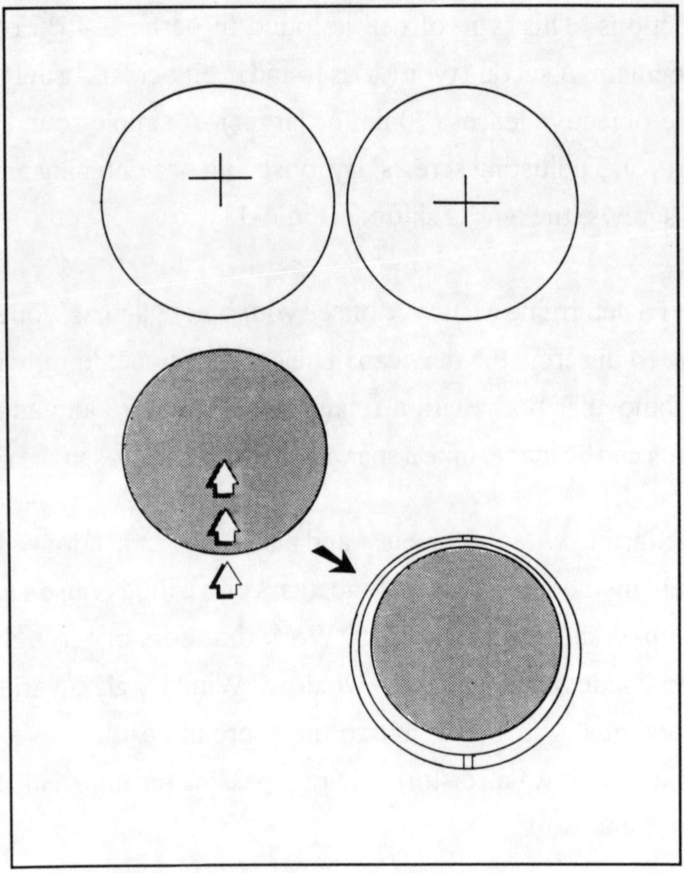

Fig. 6-12 By moving the lens center up the image through the binocular moved downward.

and the image through the binocular *down*. Go back and compare the images on the right and left sides. Did it correct the problem? Let's say it helped a little, but the right image didn't come down quite far enough even though we ran out of adjustment.

It should now become apparent that many times the collimation has slipped so far out that it's necessary to loosen and work with both objectives at the same time! In the above example, if we can quite move the image down far enough (or in reality, shift the right lens up far enough), we need to move the left lens down a bit. Make your adjustments, relax, rest your eyes and try again.

As you are making your up and down movements, it's going to be very obvious that you are also moving the image right and left. I wish this were not the case, but we are really forced to collimate in several directions at the same time, up and down and left and right. Now this may sound simply impossible for the beginning student, but after some trial and error, everything sort of falls naturally into place.

After we have gotten both sides of the binocular as close as possible, it's simply a matter of carefully tightening the retaining ring onto the objective cells. After you've tightened the rings, always go back and recheck the collimation picture. I have found that many times just the physical act of tightening the rings may shift the lenses a tiny fraction of a millimeter and throw the whole collimation job out of whack.

Let's say that you tried to move both objectives as far as they could go, but you still can't bring the images in both sides together. We have simply run out of lateral objective movement. It's time to introduce you to the "court of last resort" in binocular collimation, namely collimation by tilting the prism shelf.

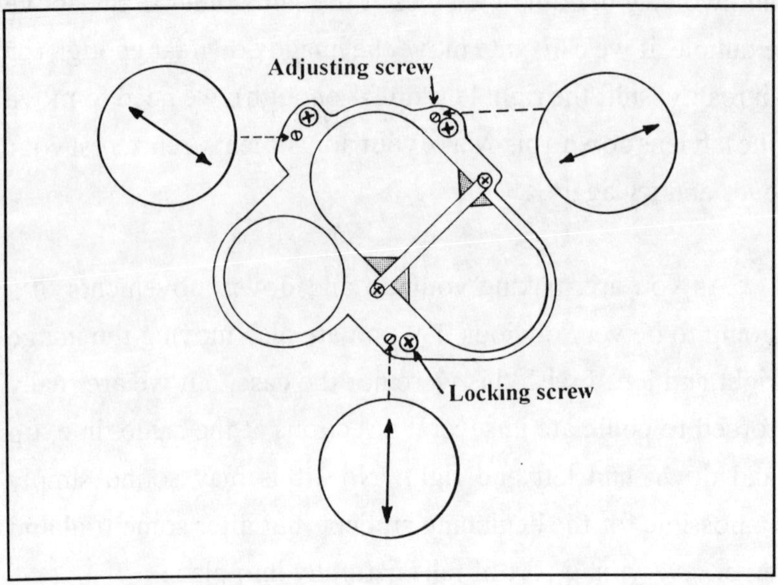

Fig 6-13 Collimation is adjusted in the direction of the arrows by three screws on the prism cluster.

This is a fairly complex procedure in that the collimation procedure must be carried out with the objectives and eye-pieces in place, but with the prism cover plates removed. Check the chapter on disassembly for the correct steps. After you have removed the eyepieces, cover plates, and ocular guide tubes, you need to screw the guide tubes back on and reinsert the eyepiece assembly back into the central focusing wheel.

Looking into the prism chamber, you will see something that looks like figure 6-13. The prism assembly is held in the binocular with three screws. Located right next to these screws are smaller adjusting screws that go through the prism shelf, but then butt up against the flange on which the shelf sets. It is suggested you try just one side of the binocular at a time as trying to adjust both sides at once can yield mass confusion!

Looking through the binocular at the test target, gently loosen the three holding screws just slightly. Perhaps 1/4 or 1/2 turn counterclockwise. You need to gently loosen the assembly without loosing complete control and having the whole prism assembly moving wildly around. Then with a small screwdriver, make the correct adjustments as outlined in figure 6-13. The direction the image will shift with each adjusting screws is indicated. The best way to look at this whole procedure is to think of it as a "tug of war" between the holding screw and the adjusting screw. Gently loosen the holding screw, then give the adjusting screw a little nudge, check the result on your target, then tighten the holding screw a tiny fraction. It shouldn't take too long to get the hang of this method. In reality, you can really do a precise job of collimation using the procedure. You've already found that collimating by moving the objectives is never perfectly precise. Even when everything looks great, that last turn on the retaining ring in the final tightening procedure always moves it just a hair in some unwanted direction. Tilting the prism shelf will usually give you enough extra adjustment room to pull the instrument into collimation. Make sure you re-tighten the larger holding screws before you replace the cover plates.

In some early production WW-II binoculars, tilting the

prisms was the only means of collimation. You will also discover some of the cheaper models being made today have no means to adjust for prism tilt or collimation by objective movement. The only recourse on these is the junk barrel.

At this point mention should be made that any collimation setting is good for only one interocular setting. If you are going to use the binocular, then collimate it at your inter- ocular distance. A really meticulous optical repairman will swear that you must collimate both at your interocular setting as well as minimum and maximum settings. Well, with all the sloppy movement inherent with mass production center focus design binoculars as well as loose tolerances in the hinge mechanism, there is no way you are going to get perfect collimation at different inter-ocular settings! You should just try and shoot for a happy medium.

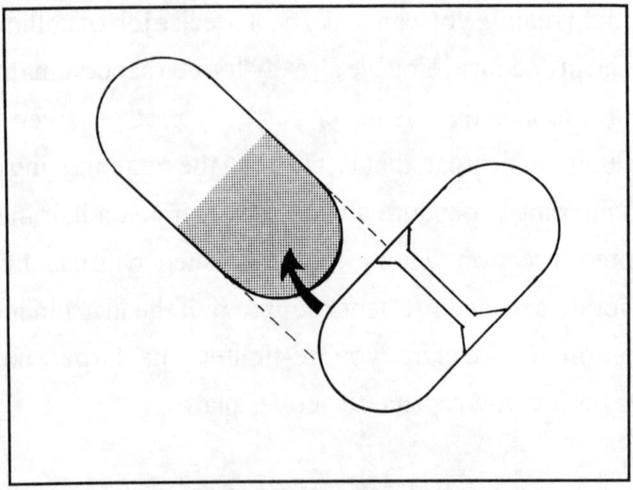

Fig. 6-14 The prisms must be exactly 90 degrees.

Squaring the prisms

We touched briefly on the subject of rotational alignment problems earlier in this chapter and stated that they were due to problems with prism squaring. Optical repairmen in the navy call this problem " lean" in that the image really appears to lean to one side or the other. Correcting this lean is always the first step they take when reassembling a binocular taken apart in their shop for repair. They have some neat little "jigs" to hold and square the prisms exactly. Since we don't run into this problem very often, it can usually be handled quite simply without much equipment. Figure 6-14 shows how the two prisms in the prism cluster on one side of a binocular fit together. Note that they must be positioned at exactly 90 degrees.

When binoculars are assembled at the factory, they are set on the prism cluster shelf , positioned exactly, and cemented with a few drops of epoxy cement. Then some type of strap is fastened over the prism and screwed down to the prism shelf.

If the binocular was subject to a hard knock , the prism may get jarred out of position. If your binocular has image lean, the first thing you want to look at is the prism shelf paying particular attention to edges of the prisms where they are glued to the shelf. If the cement is cracked, the prisms may be out of square.

To correct, we must remove both the objective lenses and the eyepieces. Review the sections in the chapter on assembly as to how this is accomplished.

Checking for prism squaring involves placing the partially disassembled binocular onto a steady holding device and looking at a plumb line or vertical straight edge. For most

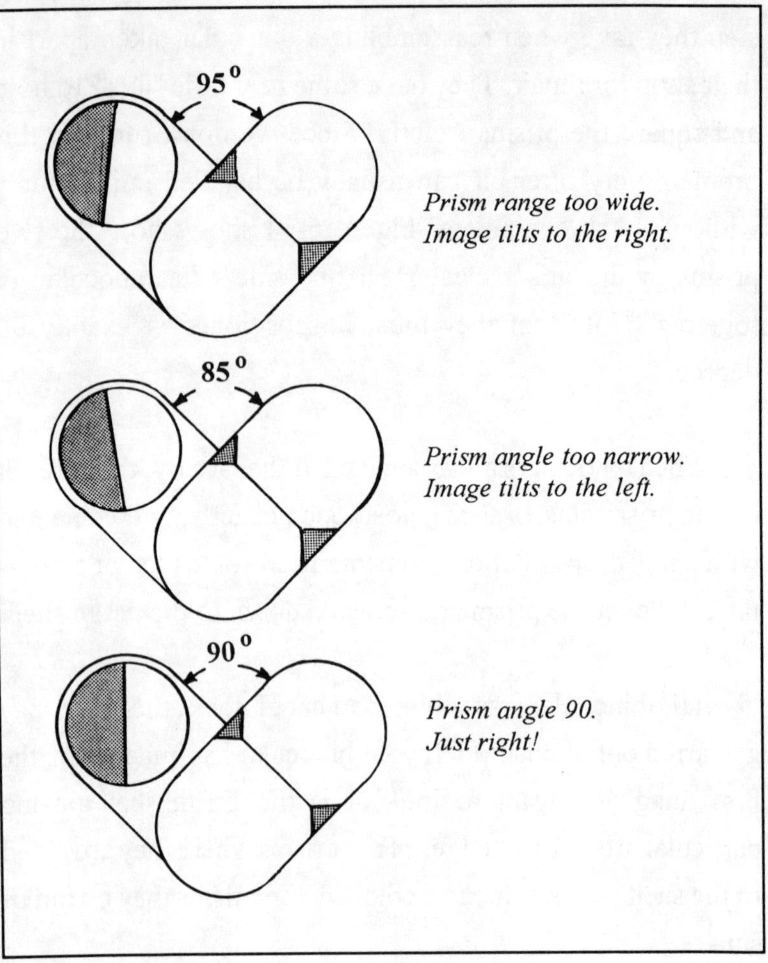

Fig. 6-15 Achieving the proper prism angle.

purposes, use the edge of a door jam. Make sure you are positioned straight in front of it.

The image through the prisms should look something like figure 6-15 if the prism is close to 90 degrees. The straight edge or plumb line will be straight up and down. Now if the prism has slipped a little to close up the angle, the plumb line will lean a little to the left. On the other hand, if the angle has opened a little, the line will appear to lean slightly to the right.

To correct, loosen the prism strap and nudge the prism a little one way or the other. When you have corrected the lean, take a tooth pick with a little 5-minute epoxy glue on the tip and tack the edge of the prism to the shelf. After this dries, refasten the metal strap and put the objectives and the eyepieces back in place.

More advanced techniques

It should be apparent that the collimation procedures discussed so far are very difficult to carry out without precise measuring devices. We can get a general idea of whether a binocular is out of collimation and even what needs to be done to correct it. However, doing a complex adjustment on the prism shelf while trying to look out a window at some nearby building is very difficult. The next step up is for the more serious student and involves collimation using some simple devices. This is really the big step separating the free hand amateur from the professional.

The Collimating Scope

First, we need a ***collimating scope***. This can be low power telescope with a crosshair reticle in the center of the field. For most of us a cheap hunting or rifle scope will do just fine. You can usually pick up a used riflescope at a gun show or yard sale for a few dollars. Stick with something in the 2.5X to 4X range with a one inch-tube. If the lenses are a little scratched or chipped, it's O.K. We just need one with a usable crosshair reticle. This collimating scope is the first element in our alignment system (Fig 6-16).

The Collimator

Next, we need a parallel light source or ***collimator***. Please do not confuse the term "collimator" with ***collimating scope***. It will become apparent shortly that their functions are quite different.

The ideal point source of light for collimating an instrument would be a star in the night sky. Some have actually suggested attempting to collimate a telescope or binocular by mounting it on some type of tripod, taking the whole thing outside and locating a bright star. The first problem you run into is trying to work in the dark. Next, we suddenly realize once again that the earth is not standing still. As soon as we locate a star and

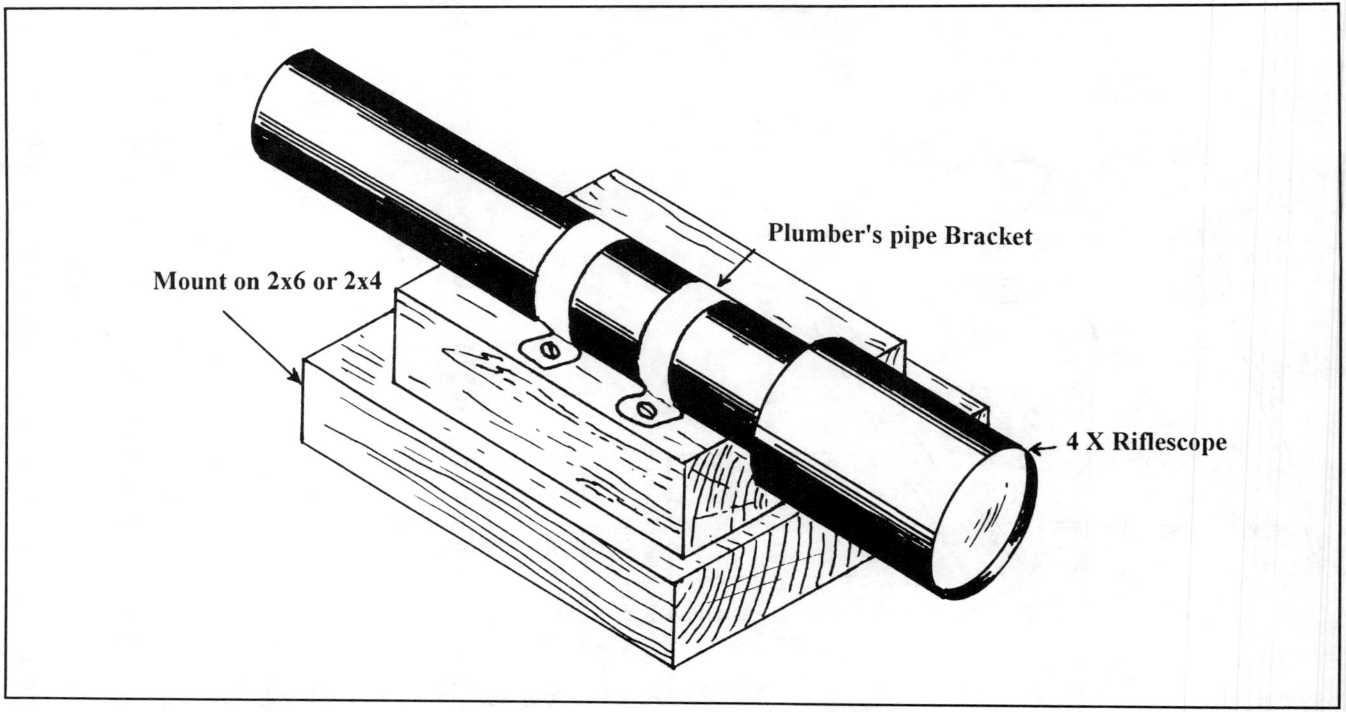

Fig. 6-16 A simple collimating scope.

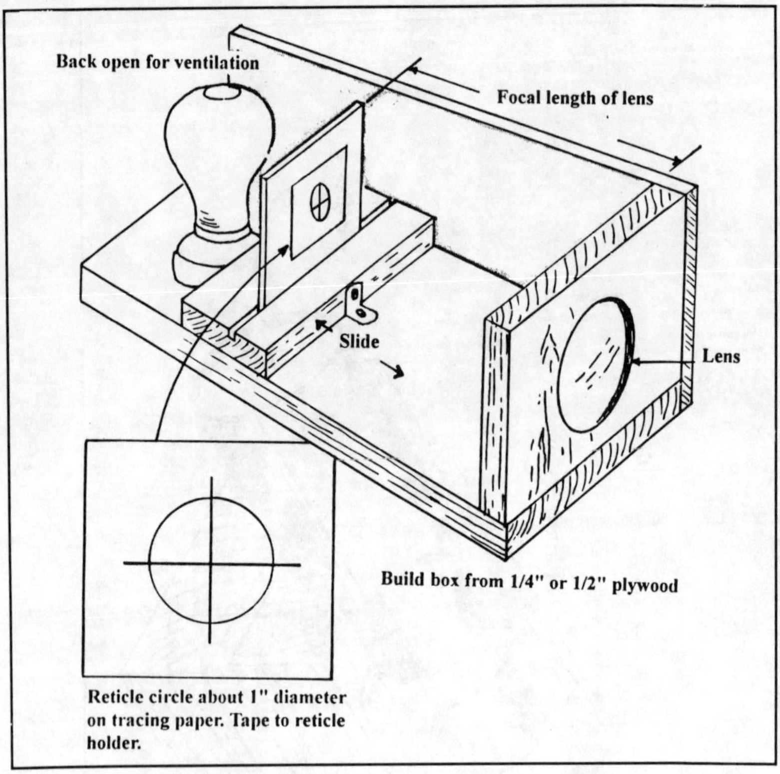

Fig. 6-17 A homemade collimator

center it in the field, it starts to slowly drift out of the field of view.

We can alleviate all of the above problems by building our own parallel light source. We need three basic components for our "do it yourself" collimator. First, we need to buy or scrounge a small lens about two inches in diameter with a focal length of about 8 to 10 inches. Check some of the sources listed in the back of this book if you can't find a satisfactory optic. I found a perfect lens for the project in an old 7x50 binocular that was being salvaged for spare parts. For our light source, we need a 15 to 25 watt electric bulb. We have to make a target on tracing paper glued to a small glass or plastic square. Figure 6-17 shows the way to lay out all the components. Remember, the distance from the back surface of the lens to the target is the focal length of that particular lens. If you don't know the focal length of a lens, it's easy to find it. On a bright day, point the lens toward a window. Bring a white piece of paper slowly in from the back until you see a sharp image of the outdoor scene come into focus on the paper. Measure the distance from the back of the lens to the paper. That will get you right in the ballpark!

Getting back to the collimator, it's best if you can build the whole thing out of wood. Most of us have enough tools and woodworking skills to accomplish this.

Now we have a ***parallel light collimator!*** That is light rays coming from the illuminated target will enter the binocular objective in a parallel manner as if the target were out there 100 yards or more.

We have our collimator, we have our collimating scope, and our binocular that needs to be tested. How are all of these things arranged? Figure 6-18 shows us what we have to do to make all these things work. The collimator is placed so it can be moved in a lateral direction without moving off axis. You can fasten a pair of straight edges to your work table so it can be easily slid back and forth.

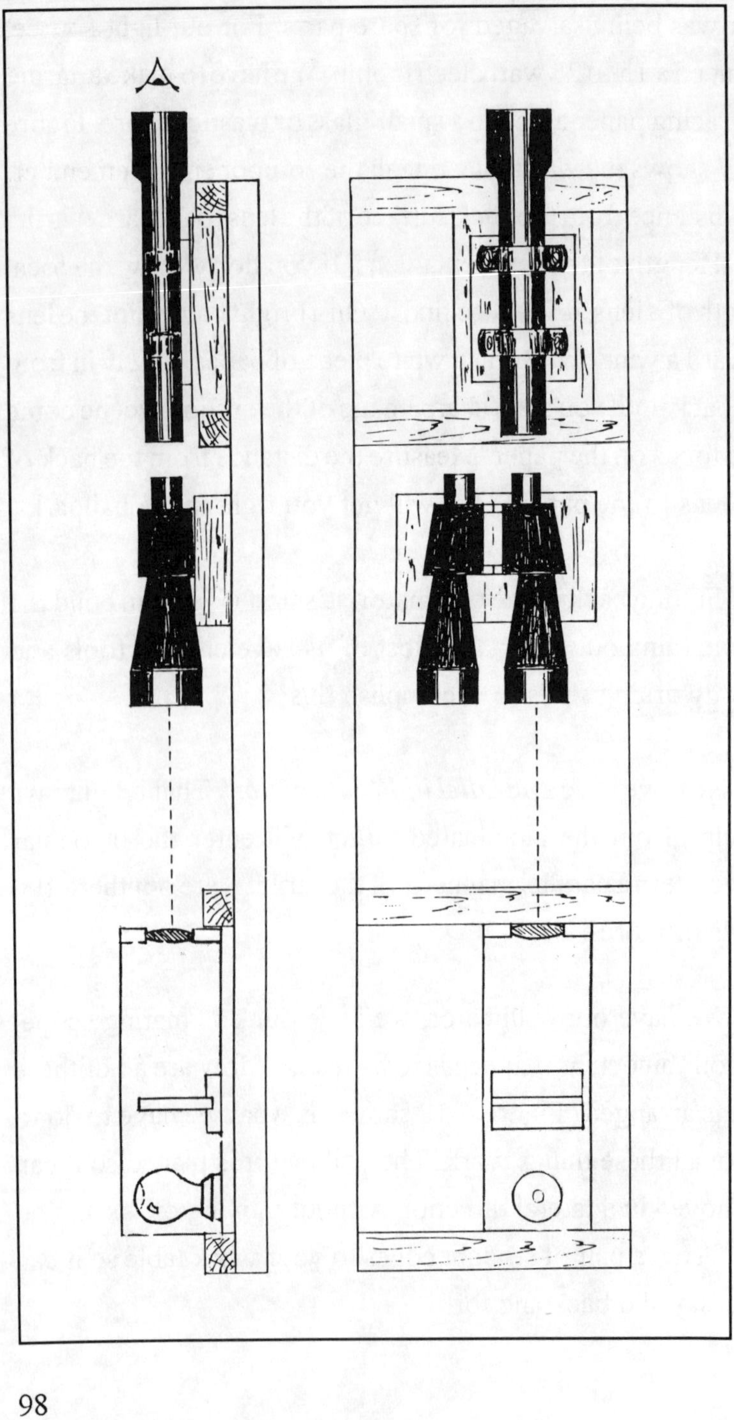

Fig.6-18 The binocular to be tested is placed between the collimator and collimating scope.

The binocular under test is mounted as securely as possible with a homemade holder. Under test we don't want this to move at all. Remember, we are also going to be working with the objectives while we try and collimate it. Mount it steady as a rock! We also want it at about the same height so the collimator lens and the binocular objective look at each other.

The collimaton scope has to be mounted in a manner similar to the collimator. That is movable between two parallel straight edges.

Now for a trial run. With the binocular removed from the setup, slide the collimation scope directly in front of the collimator. Looking through the scope you should see the crosshairs *superimposed on the target reticle* of the collimator. The height of the collimating scope may have to be adjusted slightly. Try this in several positions. With the collimator all the way to the right, and then all the way to the left. You may have to turn the focus of the collimating scope (riflescope) all the way in to bring it into sharp focus. After you have gotten used to how things look, securely mount the binocular under test into position. (Fig 6-18)

Slide the collimator directly in front of the left objective of the binocular. Slide the collimating scope in front of the left eyepiece. Sight through the scope and note the sight picture. The binocular objective is adjusted by means of the eccentric ring or tilt of the prism shelf until the reticles coincide.

Now slide the collimator and collimating scope over to the right side of the instrument. Repeat the collimation of the right objective lens. The binocular should now be in collimation for that particular interocular distance. Don't be discouraged if things don't go by the books. Practice and patience will win out!

CHAPTER 7

Cleaning Optics

The first and most important rule you can ever learn about cleaning optics be it cameras, binoculars, or telescopes is ***clean them as seldom as possible!*** Sure, you look at a lens with a little dust on it and it just plain bothers you! It sure looks like a lot of light is going to be lost and overall optical performance will surely suffer.

Well it used to bother me too until I saw a test performed by my father, a professional photographer, years ago. He took a 4x5 format still camera and shot a photo of my old VW parked about 100 feet away. Next he took a sheet of black paper and cut out five or six irregular bits of paper and stuck them all over the lens surface using a drop or two of water. It looked like there must have been 25% of the light coming through the lens blocked. An identical shot was taken of the car. What would the results be? I eagerly watched as the image in the darkroom slowly took shape in the developing tray. Comparing the two prints side by side with a magnifying glass showed no noticeable difference in the brightness or sharpness of the image! After that, a little dirt on my lenses never bothered me.

Remember, every time you rub the surface of a lens, even with the finest tissue, you are putting micro sleeks on that surface. However, from the viewpoint of a true optical repairman, when we are fixing an instrument we want to restore it to as near mint condition as we can. Part of this process involves cleaning every optical surface to absolute perfection.

There are a few basic rules about cleaning an optical surface. First, *never wipe or rub a dry lens surface with anything.* We all think glass has an extremely hard surface and in fact it does. However, modern optical instruments all have an anti-reflection coating of one type or another. when you look at a lens, these coatings are the blue, purple, and green tints visible. These coatings are also quite soft. Some of the newer *multi-coatings* are made up of 7 to 15 layers deposited on the lens surface. I have accidentally rubbed off some of these coatings through just a moment of carelessness.!

What kind of equipment do you need? It doesn't have to be expensive, but it needs to be carefully chosen. Don't use any type of towel or cloth that has been through a washing machine. City water and well water may contain micro sand particles. These sand particles get trapped in the cloth mesh and turn it into a perfect piece of sandpaper! One good alternative is to use a chemically processed tissue like Puffs. Make sure to use the pure white unscented kind. Colored dyes or perfumes added to tissue always leave smears or stains on clean lens surfaces. An outfit called Universal Photonics in Hicksville L.I. N.Y. makes a special lens tissue towel. These are perfectly pure and clean and you can really bear down on a lens without risking a scratch. These are a must if you really get serious about cleaning optics.

Next we need a cleaning fluid. Don't waste your time with camera store lens cleaning fluids. They are expensive and always leave streaks. Get hold of some industrial pure acetone. ***CAUTION-ACETONE IS HIGHLY FLAMMABLE! ALWAYS USE IT IN A WELL VENTILATED AREA. EXCESS FUMES ARE TOXIC!*** The best idea I have seen for using and storing your cleaning chemicals are little bottles sold by Edmund Scientific of Barrington N.J. They are glass, have a little pump built in and a flip-top cover.

In addition to acetone we also need ethyl alcohol. This may be a little hard to find because we need alcohol without any water mixed in. In some states you can purchase a product in your liquor store called Everclear. This is 190 proof ethyl alcohol! Excellent for cleaning glass as well as medicinal purposes.

Another possible source of ethyl alcohol is your local pharmacy. They can order 96% pure. This so called medical alcohol has a little additive mixed in with it so people don't drink the stuff trying to escape the liquor tax.

So basically your professional lens cleaning kit should consist of two pump type dispensers, one with acetone and the other with alcohol. Since these liquids are quite flammable, it's best to buy them in small quantities.

There are a few other items you will find helpful. *Dust Off* is a pressurized air can that lets you direct a burst of high pressure air at the lens surface. The same idea, but not quite effective is a rubber syringe. Both of these items can be found in a camera store. Lens tissue, Puffs and Kodak optical tissue, Q-Tips , and cotton also come in handy.

The first problem you're going to run into is how to handle a lens without leaving finger prints. The natural oil on your hand and an optical surface just don't mix. I found the best way is to wear a latex disposable glove on the hand holding the optic. Actually, it's not a bad idea to use gloves on both hands as they prevent the cleaning chemicals from coming into contact with your skin.

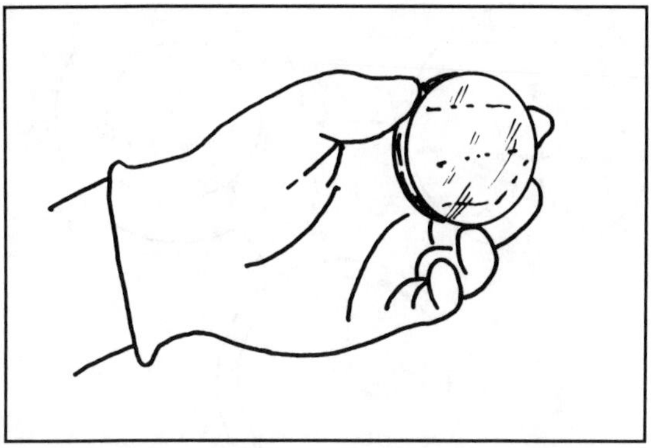

Fig. 7-1 Using a latex glove, hold the lens by its edge.

Here's the correct procedure: Figure 7-1 shows the correct way to hold a mounted or unmounted lens. Always hold them by the edge to avoid any contact with the polished surface. Blow off any loose dust particles with the bulb syringe. Dampen some lens tissue with alcohol. Make circular motions in a clock-wise direction starting at the center of the lens and moving to the edge. (Fig.7-2) You may have to repeat this process several times. Finish the cleaning with a pad moistened with acetone. You have to work fast with acetone as it is hydrophilic (loves water) so that after 6 or 7 seconds the acetone starts to draw moisture out of the air and you will streak the lens surface and have to start over again. There is no hard and fast rule on which chemical to use first, the acetone or alcohol. Try one first and then the other to see which way works best for you.

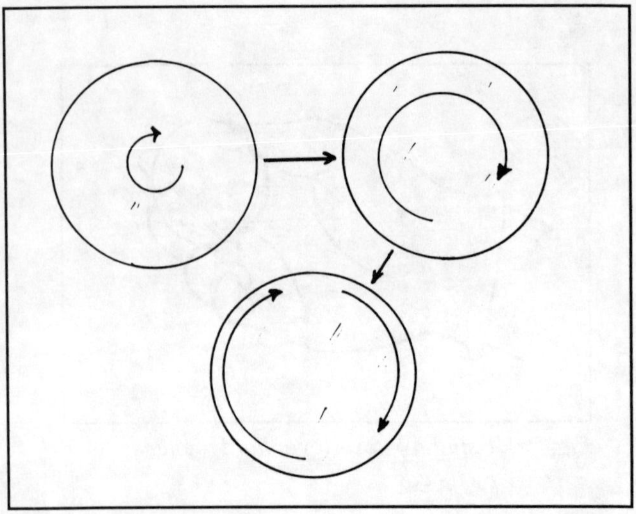

Fig. 7-2 Start in the center of the lens and make wider and wider circular sweeps.

Here is the most important part of the use of liquids on optical instruments. ***Do not apply liquids directly to the lenses.*** Moisten the tissue, then clean the lens. This is very important when cleaning optics mounted in instruments. One time I stood by in shocked silence as a camera store clerk tried to clean a customer's camera lens by first applying 4 or 5 drops onto the lens surface. Figure 7-3 is a cutaway of an eyepiece showing where this liquid goes. Remember, most lens systems are made up of several cemented elements. We don't know the long term effects of chemicals on these lens bonding cements. If you drop the liquid right on to the surface, it wicks down into the whole system! After a while, you may have a very expensive repair bill coming.

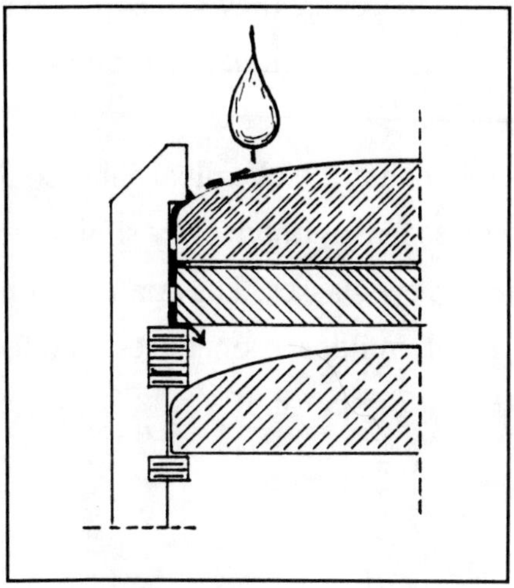

Fig. 7-3 Excess liquid migrates down into the elements.

The above cleaning methods are fine in the shop, but what do you do out in the field when your instruments become dirty? A new product has shown up in camera stores called Luminex. It is an ultra fine weave cloth that actually gets under the oil film and lifts it right off. It works very well and is a part of my accessory case for my cameras and telescopes. At a minimum, you should carry a small pack of clean tissue.

There is a new-high tech cleaning method just starting to show up on the market. It's called a "skin" cleaner. A plastic-like liquid is painted onto the lens surface with a brush. It dries like

like a coat of paint forming a pliable skin on the lens surface. You carefully lift up one edge and peel off the skin along with the dirt and oil. It's sort of like peeling your skin after a sunburn! They do a fair job. However, they are expensive and the end results are no better than traditional methods already outlined.

Please remember that the chemicals and equipment listed are for cleaning glass lenses only! They should not be used on plastic lens or plastic eyeglasses. The coatings found on plastic lenses require much milder procedures and can easily be destroyed by alcohol or acetone.

CHAPTER

8

Tools

While you don't have to have an expensive set of tools to do optical repairs, there are certain items you just can't get along without. One of these is an optical spanner wrench. This allows you to easily remove the various lens retaining rings you run across. There is a list of tool sources in the back of this book.

If you don't want to buy a spanner wrench you can easily make retainer ring tools yourself out of 1/16" sheets of steel, hard aluminum, or half-hard brass (Fig. 8-1). When making your own, it's important that they be carefully hand filed to fit the retainer ring slots they need to fit into.

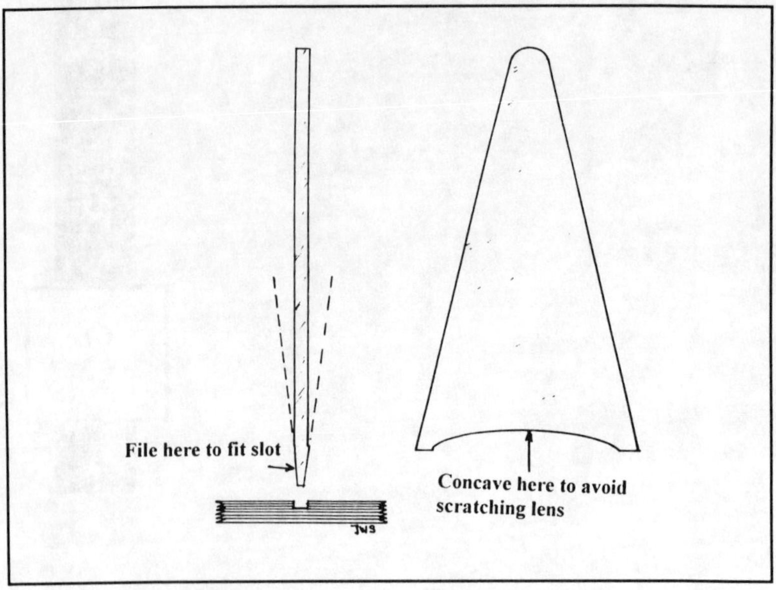

Fig 8-1 A simple homemade spanner wrench.

You're also going to need a good set of jeweler's screw-drivers. Don't skimp on these. A lot of sets coming out of Hong Kong and the Far East are too soft. After taking out a few dozen screws the blades become rounded and distorted. Those made in Japan are good. American or Swiss made are best. Make sure that the set contains several Philips-head models.

Handling glass elements is always a real problem in that you're going to constantly be getting fingerprints on the polished surfaces. You always try to handle the lens edges, but somehow a greasy fingerprint always ends up where you don't want it. Over the years I discovered the best way to avoid this is to wear latex surgeon's gloves. They prevent fingerprints and also give you a better grasping surface when handling small, smooth objects. One firm now markets special optical handling gloves that not only prevent fingerprints, they clean the optical element as you work with it!

Often times you're going to have trouble getting a good grip on a part in order to unscrew and remove it. Have a roll of black electrician's tape and some duct tape in your work kit. A stubborn part will usually yield if you put a few wraps around it to increase your gripping power.

Work Area

It should go without saying that if you are working with optics you need to develop good, clean work habits. Try and have a well lighted and correctly ventilated area set side exclusively for you optical repair. You can't do quality optical repair in a junk pile!

Over the years a can tell you how many hours I lost looking for tiny set screws that were dropped while disassembling instruments. You unscrew then, they fall and hit the surface of the table , bounce and they are gone over the edge of the table. Then you are down on your hands and knees searching the floor , looking in your pant cuffs, and in general wasting time.

Then I learned that if you *always work over a clean cloth spread on your work table*, any screws land on the cloth and stay put!

CHAPTER

9

Helpful Accessories

The first thing you learn about using binoculars for an extended period of time whether it's for surveillance, nature study, or astronomy, is that trying to hold them by hand is a real chore. After a few minutes, the muscles in the hand and arm start to fatigue and soon the image is trembling and dancing. Yes, you can reduce the problem by leaning against a tree or building edge, but sometimes this just isn't possible.

The simplest mechanical device to hold binoculars is a *tripod adapter.* These are metal devices that provide a link between the binocular and a standard camera tripod. All camera tripods have a screw stud with a 1/4-20 thread on it. Early model binocular adapters had a metal strap that came up and fastened over the binocular's center hinge pin. As binocular designs evolved towards more modern styles, the central hinge pin disappeared and manufacturers started to build the 1/4-20 female thread right into the binocular body casting.

A more common practice today is to place a threaded fitting at the distal end of the binocular hinge. They usually screw in a little plastic cap so that if you don't know it's there, it may escape you. The cap usually has a beveled, knurled edge. Unscrew it with finger pressure. (Fig. 9-1)

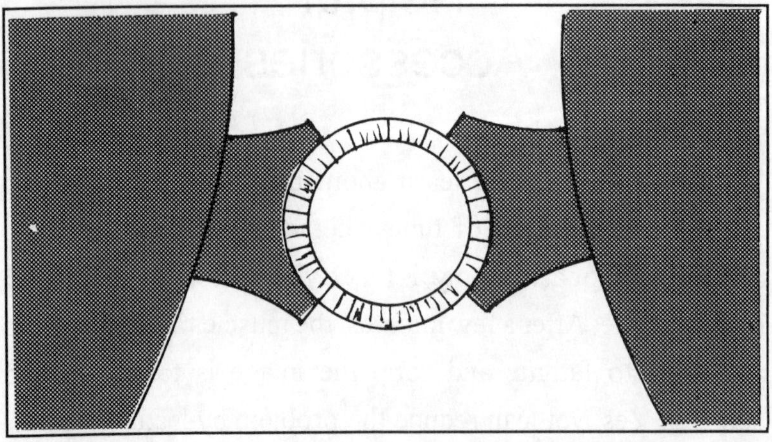

Fig. 9-1 On the far end of the central hinge, a small removable plastic cap hides the 1/4-20 mounting screw hole.

Many binocular adapters are now L-shaped brackets (Fig. 9-2) that set on top of the tripod and screw directly onto the binocular. (Fig. 9-3) The problem with mounting a binocular using this method is that you are limited in your movements to those of the camera tripod pan head. Most just go up and down and right or left and the up-down motion may be limited by the stops in the tripod.

Fig. 9-2 The L-shaped tripod adapter offers a steady mounting method that fits all camera tripods.

Fig. 9-3 A large astronomical binocular with L-tripod adapter in place.

The last few years have seen a number of ingenious devices appear on the market to give more flexibility of movement. These were outgrowths from the field of astronomy where the user needs a wide range of movement to search large areas of the night sky. Several models screw on top of camera tripods allowing a steady rest but access most any area of the heavens. (Figs. 9-4, 9-5)

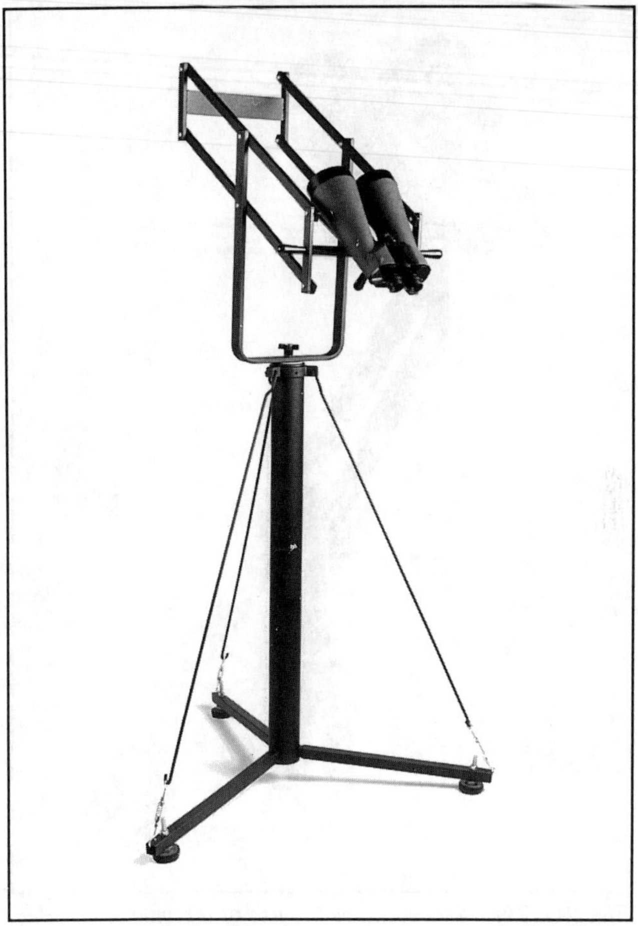

Fig. 9-4 A sturdy holder for astronomy and general sight seeing.

Remember, if the holding device relies on a camera tripod, always select the heaviest duty model you can afford. There is nothing as annoying as trying to work with a flimsy tripod!

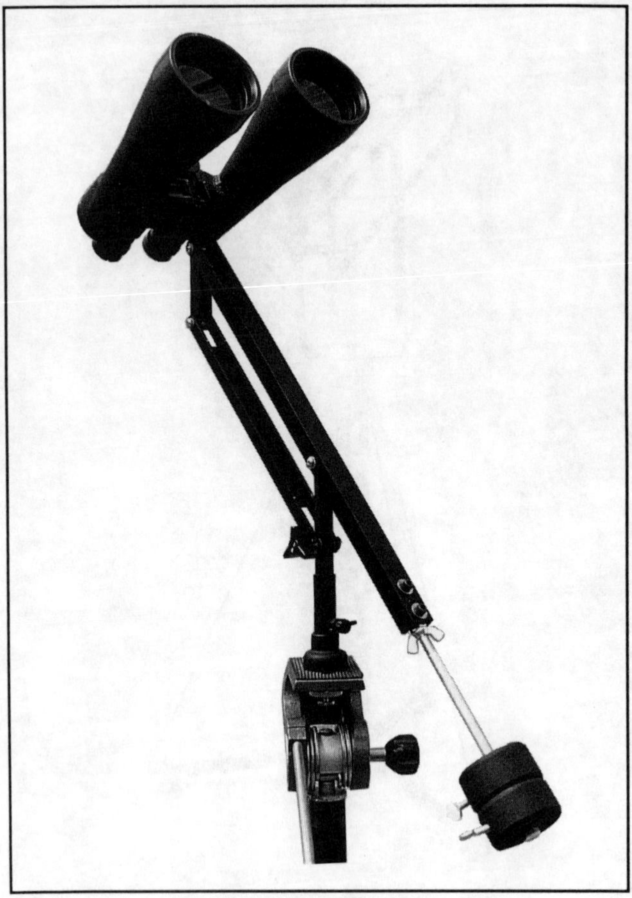

Fig. 9-5 This device allows the use of a camera tripod and curtain-weighed arm.

For a Clear View!

There are several instances where optional glass filters can give you an improved view. Let's say you use your binoculars for boating and fog and haze are constant problems. By screwing a pair of glass filters over the eyepieces, usually #12 screwing a pair of glass filters over the eyepieces, usually #12 or #15 yellow, contrast and visibility are improved. It's sort of like adding yellow filters to your auto headlights!

Since it may be difficult to find glass binocular filters at your dealer, you can accomplish the same thing by cutting round circles of color gelatin camera filter material and then taping them over the objective or eyepiece. Many different colors are available at your camera store

In astronomy, the glare of city lights can really spoil the view. Special filters that cut our certain light bands, sodium and mercury for example, can visibly darken the sky and improve nighttime views of the heavens. These filters are called LPR or light pollution filters.

Keep those lenses dry!

If you use a binocular at night, you quickly discover that as the night temperature cools, dew suddenly settles on your equipment. Dew on a lens surface will destroy your view quicker than anything I can think of! For some of the larger binoculars there are slip on dew caps that extend out in front of the lenses several inches and greatly reduce the problem (Fig. 9-6). You can also make your own if you have smaller binoculars by making cylinders out of heavy art board and them taping them on the objective housings.

Fig 9-6 At night, heavy dewing of the objective can be avoided with dew caps.

snow or water, a pair of polarizing or neutral density filters can really cut the glare.

The Sun - Be Careful !

It goes without saying that any family member having access to binoculars should be warned about ***the dangers of either accidentally or purposely pointing the instrument directly at the sun.*** The tremendous concentration of the sun's rays by the objective lenses can result in instantaneous and permanent damage to the eye's inner structure.

Adults pursuing a scientific hobby like astronomy may , with special filtering devices fitted over the objectives, safely view the sun for short periods to observe sun spots or solar eclipses. Several firms make this specialized equipment. (Fig. 9-7)

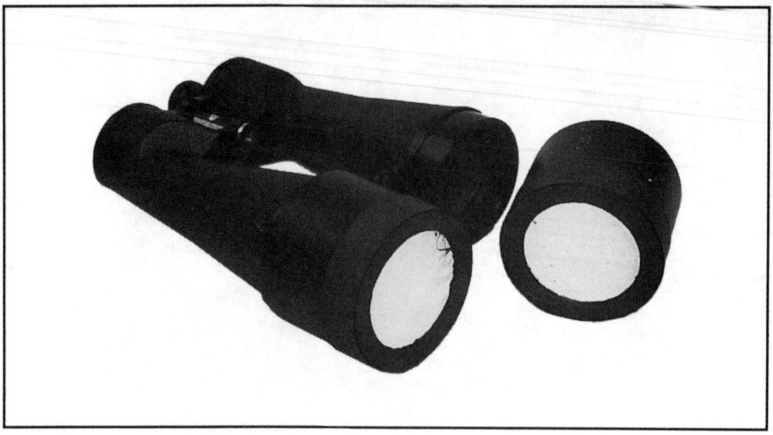

Fig. 9-7 Special filters allow binocular users to view solar eclipses.

Proper Neck Straps

The neck straps that come as standard equipment on most binoculars leave a lot to be desired. Most are about 1/4" wide, made from vinyl and if your instrument has any heft to it, you find the strap digging into your neck. Visit your local camera store and look at the wide array of auxiliary neck straps available. The best ones are woven nylon, which is stronger than vinyl, and have a wide support band where the strap touches your neck. *Never use a binocular or let anyone else use* your binocular without the neck strap in use. A binocular repairman's best friend is a missing strap.

CHAPTER 10

Repairing Separated Lenses

You're more likely to encounter a lens that has become separated than you are to find a cracked or broken one. Lenses are normally mounted in and thereby protected by metal cells. It takes some real trauma to actually damage the glass itself.

Most older binoculars used Canadian balsam to cement compound lenses together. Canadian balsam is really nothing more than a very refined version of the sap you see weeping out of pine trees in the springtime. It just happens to have an index of refraction very close to glass so it makes a perfect optical cement for joining optical elements together. Over the years, moisture and ultraviolet light slowly breaks down this balsam and you have what is known as a *lens separation.*

How can you recognize a separated lens? At first glance, you may think you are looking at a broken or fractured element. The lens looks like it has a number of concentric rings spreading out from the center to the edge. It's almost as if there was a multi-colored rainbow between the elements. A broken lens on the other hand will have a more jagged look to it. There may also be tiny slivers of glass present. If the separation has become complete, the lens takes on a "milky" opaque cast. Figures 10-1A, 1B, and 1C show us what some of these common defects look like.

Review the section on taking out the lens with its retaining cell. After the lens is out of the cell, we have to get the two elements apart. The first method to try is to soak the lens in acetone. Handling acetone always requires some extra precautions due to its toxic properties. Make sure you always have *adequate ventilation*. When possible work out of doors! If you have a garage, open the door and work in there.

Use a glass jar with a screw top. First, we need to mark the side of the lens element with a pencil so we know how it goes back together. Put a little (>) arrow right on the side of the

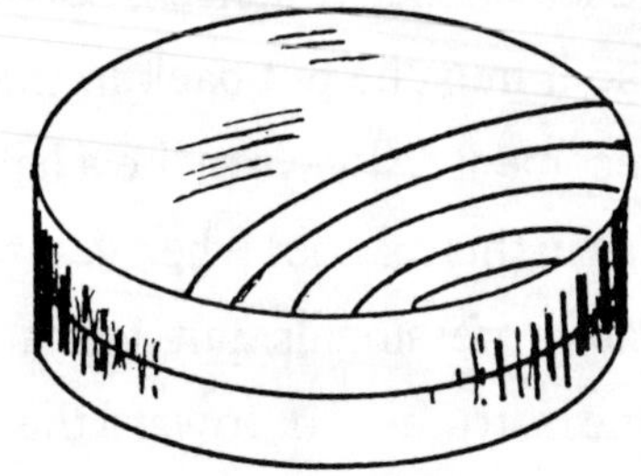

A separation of lens elements appears as a rainbow between the elements.

With age, the cement between the elements begins to show odd shapes and designs.

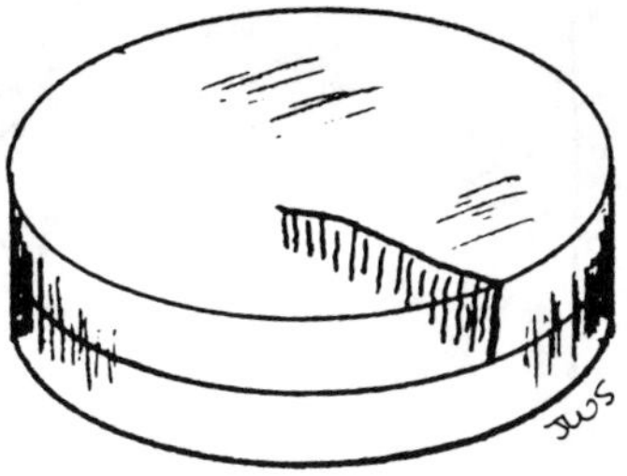

An example of a glass fracture or crack in one element of a cemented two element lens.

Fig. 10-1A, 10-1B, 10-1C Separated and broken lenses can have similar appearances.

elements with a sharp lead pencil .Each lens surface has its own radius of curvature so it must be put back in exactly the same position when we recement. Always make a habit of immediately marking any lens in this manner when you remove it from an instrument. Most professionals put the arrow pointing toward the front of the instrument or toward the object you are viewing .(Fig. 10-2) When you put the lens back together, just realign the arrow. Getting back to the acetone soak, sometimes it takes two or three days before they come apart.When the lenses are ready to come apart, they usually look rather white and milky. Press the lens a little sideways to finally break them apart. You can put quite a bit of pressure on them without fear of breaking them.

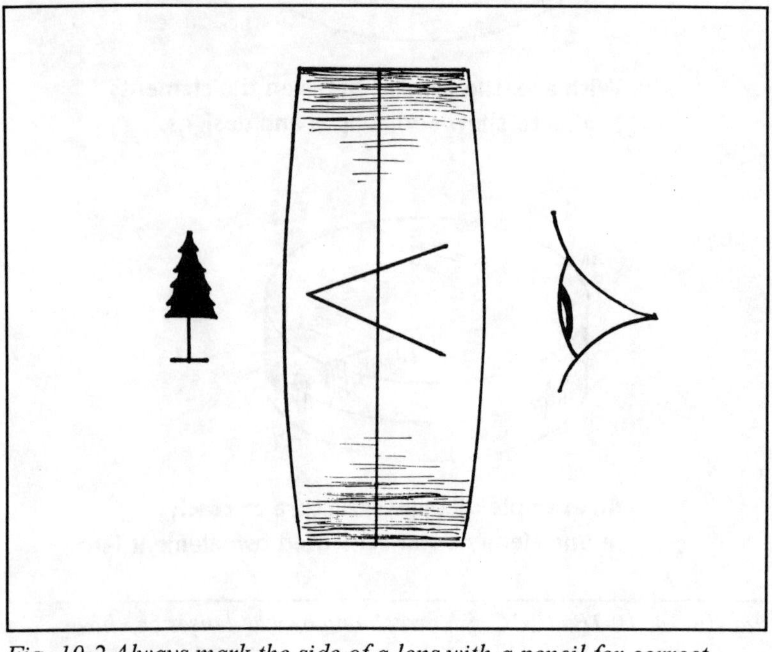

Fig. 10-2 Always mark the side of a lens with a pencil for correct orientation.

You may come across a really tough lens that will not come apart using the acetone soak bath. The folks who repair instruments for the Navy tell us that they have found a sure fire method. Put the stubborn element in a Pyrex baking dish and put it in a 300 degree oven for 30 minutes. Take it out, let it cool slowly ,and try the acetone soak again.

After the lens is apart, very carefully clean it with alcohol and acetone to remove all traces of the old cement. *This is one place where you really have to do a good job !* These lens surfaces have to be sparkling clean. If they are not perfectly clean and you re-cement them with any dirt or lint specks between the glass, you end up with a second class job!

Re-cementing the elements

You could re-cement the elements with Canadian balsam. You're better off using one of the new ultraviolet cure cements now on the market. You'll find a listing where to buy them in the back of this book. The great thing about the new cements is that after you have correctly positioned the elements, just put it in a sunny window and in 15 minutes it has set perfectly!

Re-cement the elements in the following manner: With one lens on the table setting on a clean piece of paper, place a *few drops* of cement in the center. Press the second lens on top and gently rotate the lens back and forth with a little pressure in sort of a circular rocking motion. By natural action, you'll see the lens cement spread out between the elements and work its way toward the outer edge.

We want a perfect sandwich of lens cement between these tow elements. Don't panic if you observe lots of tiny bubbles. With gentle pressure and constant rotation, these will work their way to the edge and disappear. The end result should be ***perfectly clear with no bubbles.*** You may have to do the whole thing over several times to get it right.

O.K. , now you have the lenses back together, but remember the cement takes time to cure. Most of the newer cements will set in 30 or 40 minutes. We have to hold these elements together while the cement sets. If they are not held exactly in position with the sides absolutely parallel, after a few seconds the top lens will slide slightly off-center. If it dries in an off-center position and you reassemble the binocular, it will be suffering from what is termed an "off-axis optical system".

I know many people are impressed when they hear a camera lens or telescope eyepiece has a seven or eight lens system.Just remember how difficult it is to cement just two lenese together. Think of the problems inherent in cementing a multiple lens sandwich! To me, this helps explain why a pure simple design often performs best.

How can we hold these lenses while the critical drying period is taking place? The way the optical repair experts do it is with a set of "V blocks". If you don't have any V blocks, you can do the same thing with a set of three blocks. Slide the blocks against the edge of the lens from three points about 120 degrees apart and leave them there until the cement sets.

Another method that can be used requires the elements to be wrapped with tape. Just use some black electrical tape or duct tape.

You'll find that this whole lens cementing process can become quite messy as the excess cement oozes out from between the lens sandwich. Don't worry about it. After the cement dries, you can easily clean all this off using acetone. The non-optical sides of the lens can have the excess cement removed by an acetone wipe. After the lens is clean and dry, it can be put back in its cell and replaced in the binocular.

Can you anti-reflection coat a lens?

Unless you are preparing binoculars that were made prior to WW-II, at some point you'll run into the problem of damaged or missing anti-reflective lens coating. All modern optical instruments have these coatings to improve light transmission. While the coating is rather durable, normal wear along with improper lens cleaning methods may damage it. Can a lens be re-coated? Yes. Is it 100% necessary? Probably not.

The problem we run into is that commercial coating is usually done in large lots. First you have to find a laboratory willing to take a single lens and then cost is a big factor. A minimum "start up" on a coating chamber may run as high as $100!

You have to realize that optical coating is done a high temperatures in a vacuum chamber. This process will usually destroy the cement holding lens elements together. So to do it right, you have to take the lens apart and then re-cement again after coating.

In practical terms, you may be losing only 7% to 8% in total light transmission by having one lens uncoated in a binocular Hardly of any significance considering the trouble and cost involved.

A little discussion here on lens coatings may be in order. Up until WW-II, lenses on binoculars and cameras simply were not coated. The same held true with large professional astronomical telescopes. If we go back now years later and compare the performance of these old uncoated instruments with modern units with all their advanced coatings, the surprising thing we find it there is not a lot of difference in overall image quality. Sure, advances in technology are fine to a point, but don't ever get the idea they are a substitute for basic optical quality. I have examined modern fully-coated binoculars of inferior optical design and quality that could not hold a candle to some Zeiss units made in the 1930's. Don't get all upset if one or two optical surfaces in your binocular have thin, slightly damaged, or missing optical coatings. Believe me, you are never going to know the difference!

CHAPTER 11

Binocular Mathematics

When it comes to binocular terminology, sometimes things get a little confusing. On one hand, the diameter of the lens is always expressed in millimeters (there are about 25mm to every inch) yet on the very same instrument the field of view may be expressed as so many feet at 1000 yds!

Finding the exit pupil and field of view are fairly simple calculations that you may want to commit to memory especially when purchasing a new instrument.

Finding the exit pupil

Exit Pupil = Objective Diameter (mm) / Power

As an example, let's take a typical 7X50

$$\frac{50}{7} = 7.1 \text{ mm exit pupil}$$

Field of view

Linear Field [ft/1000yd] = Angular Field [deg] × 52.5

Apparent Field = Angular Field × Power

Resolution

Resolution [arc sec] = 116/Objective Diameter [mm]

The resolving power of the human eye (corrected) is about 60 seconds of arc. Let's take an example of a binocular with a 30mm lens. This would yield a theoretical resolution of 3.9 arc seconds. Yet a 30mm binocular is considered excellent if it passes a military specification chart test in the field at 4 arc seconds! Things really get bizarre when we move up to 100mm. This would yield a theoretical figure of 1.2 arc seconds! I have seen expensive astronomical telescopes in the 100mm size range that can not really yield this performance level! *The conclusion is that manufacturer's specifications regarding resolution of their equipment tend to be theoretical at best.*

Relative brightness and twilight factor

Relative brightness and twilight factor are indices that in the past many authors spent a great deal of time and effort carefully explaining. I have included the mathematical formulae for historical information only.

Please realize that these terms evolved long before advances in multi-coating technology when many units being sold were only partially anti-reflective coated. They also don't take into account recent revelations about the effect of age on the human eye. Their inclusion in manufacturer's literature and binocular comparison charts is of dubious value.

Relative brightness

Relative brightness is the relationship between aperture and magnification.

$$\text{Relative Brightness} = (\text{Exit Pupil size})^2$$

A 10x70 would have a relative brightness of 49.

Twilight Factor

Theoretically, the Twilight Factor measures the sharpness of the image in adverse (low) light conditions. Mathematically

$$\sqrt{\text{power} \times \underset{\text{(in mm)}}{\text{objective lens diameter}}}$$

it is determined by the following formula.

Conclusion

From all of the formulae presented in this section, several points become clear. Within reason, *the larger the objective size and the resulting increase in pupil size, the brighter the image. The better the quality of the optical coatings (Multi layered) and prisms (BAK-4), the brighter the image. Finally, under certain low light conditions,(dawn and dusk) a little more magnification usually gives a little more image detail.*

CHAPTER 12

A Brief Personal History

My first recollection of any interest in the science of optics occurred in 1946 just after the close of World War II. At that time, Ann Arbor Michigan, besides being noted for its great university, was home base for a company called Argus Optics. Argus supplied the government with a variety of lenses and prisms for military use. Rumor had it that most of the optics were going into the tens of thousands of pair of military binoculars built during the war. Argus Optics proudly represented our city's contribution to the final victory.

The Argus plant was a sprawling complex of red brick buildings located on the city's west side. As it so happened, my old elementary school butted up against the factory on one side. There behind a formidable cyclone fence lay one of the shop's loading docks. A veritable treasure cove of forbidden delights to any 10- year old boy. Glistening in the sun, brighter than any diamonds, were barrels and boxes of rejected scrap lenses awaiting pick-up for recycling or destruction.

With the war over, security at the plant relaxed and some of the more athletic among us were able to scale the fence near the gate and raid the scrap barrels without any dire consequences. The rewards were beyond our wildest dreams! Lenses of various sizes, perfect in every respect to our eyes, were sorted and arranged on carefully laid out handkerchiefs. Everyone became an expert " horse trader ". The big 2 inch ones were the most prized being worth at least 4 or 5 smaller ones plus a giant colored "agie" marble thrown in for good measure.

It was only a matter of time before our first crude experiments began using some of our liberated lenses. The most awesome discovery was that by focusing the rays of the sun on some bits of paper or dry leaves held at the exact focal point of the lens, soon a tiny tongue of flame flickered to life! On sunny days, many of our teachers' peaceful lunch breaks were rudely interrupted by having to dash outside and stomp out small grass fires that mysteriously appeared on the playground.

By gluing some of the larger lenses in one end of a cardboard tube and then holding some of the smaller ones at the other end, lo and behold, a fuzzy magnified image appeared! Little did we know then that we were repeating some of the very early experiments of Galileo, one of the world's greatest scientists, over 350 years ago. Galileo's first telescope consisted of two lenses and a section of organ pipe for the main tube.

Over the next year the quality of our homemade instruments gradually improved to the point that some of our nightly "recon" missions were planned and carried out chiefly near the houses of some of the neighborhood girls who had suddenly blossomed by the end of the sixth grade.

Several years later my uncle, a confirmed bachelor who worked as an inspector in a precision ball-bearing plant, showed up at our house one day with a telescope he had assembled from parts he purchased from a company out east called Jaegers. And what an impressive instrument it was! With its 4-inch objective lens in a polished aluminum tube, it was nearly 5 feet long. It had several eyepieces you could change for different powers. While other people were struggling to raise families and meet mortgage payments, my uncle always had the extra bucks to buy the really neat stuff most of us could only dream about.

Taking his 4- inch telescope out behind our house and up a small hill, we could look out across the Huron River Valley. About a mile and a half away were some old barns and a farm house. With the right eyepiece you could actually read the license plate on an old Ford pickup truck parked near the barn. People were walking around doing farm chores completely oblivious to the fact that they had "visitors" from across the valley. Through the magic of optics, we had projected ourselves through space into these strangers' world over a mile away! It's no wonder that during the Dark Ages, optics were looked upon as sorcerer's tricks and the work of the devil. From that time forward, I was never to lose my fascination with the science of optics!

During high school, the instruments my friends and I made grew larger and more complex. This was an exciting time for our nation. We were locked in a race for space with the USSR. Our schools expended their programs where science was concerned. About the time our astronauts were making their first suborbital flights, I was busy grinding and polishing an 8" mirror for an astronomical telescope science fair project. Many an enjoyable evening was spent watching for satellites pass overhead or studying the details of Mars and the craters on the moon.

Fade out and refocus again in 1966 in a place called the DMZ (demilitarized zone) in a country called Korea. This time the trip was courtesy of the U.S. Army. The Second Division sat eyeball to eyeball with some tough turkeys in North Korea across a barbed wire no man's land. Tensions were high and firefights were common as soon as the sun slipped below the horizon.

I headed a small detachment working out of the 43rd MASH hospital about 45 miles north of Seoul. We were responsible for the procurement of a sizable part of the fresh fruits and vegetables our troops used. These were all obtained from local Korean farms. In order to insure that Uncle Sam's sanitary regulations were followed and no " night soil " was used, a small party, perhaps one or two of us along with a Korean national would walk around those nasty little hills the Koreans call mountains.

The areas we inspected were remote, we were lightly armed, usually with nothing more than a G.I. issue .45 , and there were constant reports of heavily armed North Korean infiltrators moving through our sector. Usually disguised as farmers and wood cutters, they were hard to spot. After about a week on these patrols, the first time I got back to the PX, I picked up the best pair of binoculars I could find. They happened to be a set of Nikon 8x30s. Lightweight and sharp as the devil, they became my constant companion for the rest of my tour.

The Second Division had a number of tall wooden observation towers and it was here I got my first hands on use of a pair of giant 20x125mm observation binoculars. They equaled the power and size of the so-called " battleship" binoculars found on many naval vessels. I was simply fascinated by the views through these monsters and noted they were of Japanese manufacture.

When time came for a few days of R & R, I ended up in Japan tracking down the maker of these big 20x125's. These visits to Japan opened up a whole series of contacts within the Japanese optical industry that I maintain even to this day. In 1966 many of the high-tech industries in Japan were still in their formative years. Coming from an industrial area like Michigan, I was really shocked at the lack of modern plants and facilities. There were hardly any brick or steel buildings. Lens makers and optical instrument assemblers worked out of fragile one and two story wood buildings scattered throughout the Tokyo area. No matter how spartan the surroundings, these were people dedicated to the principles of hard work and constant quality improvement. These very characteristics would in the years to follow allow these people to completely dominate most of the world's precision industries.

CHAPTER

13

Export Standards

The story of Japanese quality control and their government's export standards have relevance for us in that whenever we purchase a product whether it's an automobile , electronic device, or a binocular, the more we know about it and understand it, we can ultimately make a better purchasing decision. If we are trying to repair a binocular, it's always a good idea to review the inspection standards material.

The only regulation offered for our protection on imported optical equipment is the U.S. Custom's Service requirement that products be identified as to country of origin. This regulation obviously has nothing to do with the quality of the item and over the years some unscrupulous businessmen have found ways around this by shipping items from China, Hong Kong ,and Korea to certain European countries where they are labeled as having been manufactured in that country.

It's estimated that over the past 20 years more than 90% of the binoculars sold in this country were of Japanese origin. A basic understanding of binocular quality and construction must begin with the quality standards established by the Japan Telescope Inspection Institute.

The institute was established in the early 1950's to control and maintain the quality of optical goods being exported throughout the world by the rapidly expanding optical industry. American photojournalists covering the Korean War discovered the excellent Nikon photo equipment and suddenly everyone wanted them.

About this same time, an expert in quality control named W.E. Demming was making the rounds lecturing Japanese industry and government officials on the value a strict program of inspection and rejection of defective parts. The optical industry was one of the first to fully embrace his concepts.

Scattered throughout Tokyo are a number of drab, gray two story buildings bearing the seal of the Japan Telescope Inspection Institute. While the name would imply their main concern is telescopes, in fact their inspection area covers binoculars, astronomical telescopes, microscopes, and related optical equipment.

All day a steady stream of small vans and trucks arrive from contractors carrying production lots of optical equipment for testing and certification needed to obtain the small oblong gold " PASSED" seal we have seen on optical items over the years.

For inspection purposes, Japan is divided into geographic zones. A number "1" could simply mean that station was located in a certain part of Tokyo and has no relation to the quality of the instrument. In 1985 , the JTII added an additional gold seal designated E.I.S. standing for Exceptional Inspection Standard. This may be found on some of the more expensive optical items.

The following is a translation from the original Japanese:

Japanese Export Inspection Standard For Prism Binoculars.

(Approved by the Ministry of Trade & Industry)

Scope- This specification covers standard requirements as applied to binoculars using Porro or Roof prisms as the erecting optical system.

Appearance- Optical Components: All components shall have no cracks or chips, fog or mildew, and cemented optical surfaces shall be free from feathering, scratches, dirt or other imperfections or impurities which may effect their optical qualities or appearances.

Other Components- Metal parts shall be free from rust, corrosion and breakage and will be free from scratches, uneven surfaces and other imperfections that may effect their use or appearance. Leather or plastic covering over the binocular shall adhere to the surface and not readily peel off. Steel material shall be subjected to satisfactory anti-corrosion treatment. Binoculars shall be distinctly marked on a part of the body with magnification, effective diameter of the objective, real field of view, maker's mark or identification number. These marks shall not easily fade away. In the case of binoculars designated WIDE ANGLE, the normal field of view shall not be less than 65 degrees.

Materials- All component parts shall be made of appropriate materials for strength.

Construction and Function- The adjusting range for the interpupillary distance shall have an adjustment range of 60 to 70mm or more. The adjusting device of diopters shall have an adjustment range of + 2 to - 4 or more. In the case of center focus design, the adjustable right-hand ocular shall have an adjustment range of +2 to -4 or more. The eye-relief shall be 4mm or more from the outer surface of the eye lens.

Anti-Reflection Treatment- 1) In the case of binoculars marked COATED, at least the inside or outside surface of the objective or the inside or outside surface of the eyepiece shall be subjected to anti-reflective treatment. 2) In the case of binoculars marked FULLY COATED, all air-to-glass surfaces of all lenses or prisms shall be subjected to anti-reflective treatment. 3) In the case of binoculars marked MULTI-COATED, at lease the outside or inside surface of the objective or the outside or inside surface of the eyepiece shall be subjected to anti-reflective coatings of 5 layers or more. 4) In the case of binoculars marked FULLY MULTI-COATED, all air to glass surfaces including the prisms shall be subjected to anti-reflective coatings of 5 layers or more. In the case of binoculars marked UV COATED, the objective lens shall have anti-reflective coatings on all 4 surfaces. That is they shall be coated before the cementing process.

Prism Materials- Prism material shall be of two types: BK-7 or BAK-4. In the case of BAK-4 , this material shall be so designated on the outside of the body in a non-fading marking. (variance) Binoculars containing BAK-4 prism material may at the assembler's option be marked BP-2UV on the cover plates.

Prism Setting- Prisms shall be fastened to the prism shelf or platform within the binocular by means of metal straps, metal straps and cement, or metal straps and set screws. In no case will cement only be an acceptable means of fastening prisms.

Miscellaneous- Dimming of the image at the edge of the field stop, inequality in size between right and left visual field, and other defects of a similar nature shall not be appreciable. The inner surface of the binocular shall not reflect stray rays of light or exhibit rough or poorly machined surfaces. The construction and function of the adjusting device for interpupillary distance, diopter and other movable parts shall be accurate and smooth under ordinary service conditions.

Aberrations- When a test target placed at infinity is brought into sharp focus, the objects located near the center of the visual field shall have as little aberration as possible. The circumference of the visual field of view shall be free from excessive aberrations.

Magnification- When binoculars are focusing on an infinity target, the permitted limit shall be + or - 5% of the normal magnification.

Aperture of the Objective Lens- The permitted limit shall be + or - 1mm.

Diameter of the Exit Pupil- When binoculars are focused

on an infinity target, the permitted limit shall be + or - 10% of the normal diameter of the exit pupil.

Real Field of View- When binoculars are focused on an infinity target, the permitted limit shall be + or - 5% of the real field of view.

Collimation- (Alignment) When two parallel rays of light are projected into the two objectives, the conjugate rays emerging from the two eyepieces shall be parallel with the following limits (expressed by the value of the real angle in minutes 0 at the infinity focus of the eyepiece for any setting of the interpupillary distance between 60 and 70 mm as shown in the following chart:

Magnification	Direction	Collimation (in min.)
Under 4.5X	Vertical	8 and under
	Divergent	20 and under
	Convergent	15 and under
4.5 to 10X	Vertical	4 and under
	Divergent	10 and under
	Convergent	6 and under
11 to 15 X	Vertical	3 and under
	Divergent	8 and under
	Convergent	4 and under

15 to 20X	Vertical	2 and under
	Divergent	6 and under
	Convergent	3 and under

Resolution- The resolution of each half of the binocular is measured by using a parallel bar test target with black and white stripes of equal width and expressed in seconds of arc. See following chart for limits based on lens diameter.

Objective Diameter	Center of Field	2/3 of Field
20mm	6 arc. sec.	15 arc sec.
30mm	4.2	13
35mm	3.5	12
40mm	3	10.5
50mm	2.5	10
60mm	2.0	9.5
70mm	1.8	8
80mm	1.5	7

Difference in Magnification Rt. vs. Left Side-Magnification of right and left sides of the optical system shall not differ from each other by more that 1.5%

Image Tilt- (Squareness of the prisms) The image of a plumb line formed by the optical systems shall be vertical within 2 degrees and parallel within 40 min.

Change in diopter in Zoom models- In the case of variable power models, the change in diopter shall not exceed 1 diopter when focused on an infinity target located at the center of the field of view and zoomed from minimum to maximum settings.

Parallax- (in models with reticules) Parallax between the image of an infinity target and the reticule graduation nearest the center of the field shall not exceed + or - 0.5 diopter.

Anti-Vibration characteristics- There shall be no visible damage or loosening of assembled parts as a result of vibration corresponding to the falling action at a constant frequency of 250 times per minute with an amplitude of 4mm. for a period of 5 minutes.

Anti-Pressure characteristics- Binoculars marked as WATERPROOF construction shall undergo an internal pressure check of 0.1 Kg/cm. for a period of 5 minutes. The loss of pressure shall not exceed 10%.

Sample testing Methods. *Lot sizes, acceptance and rejection numbers are beyond the scope and intent of this book. Those interested in further information are encouraged to consult a text on sampling and quality control methods.*

CHAPTER

14

Glossary of Binocular Terms

Abbe Prism- Also know as a roof prism. Used in modern binoculars to invert the optical image and correctly orient the view.

Aberration- All those errors, either from design or manufacturing techniques that result in an imperfect image,

Accommodation- the ability of the eye to fine focus both near and far by eye muscles holding the lens. Diminishes as we age.

Achromatic Lens- A lens having two or more elements usually of differing materials. An example being crown and flint glass.

Antifog Coating- A chemical coating applied to a lens surface that prevents the condensation of moisture.

Aperture- In an optical instrument, the maximum clear opening of the front objective lens. Also referred to as"clear aperture".

Air-Spaced- referring to binocular objectives in which the two-element glass lens is separated by thin foil or metal spacers. Allows better resolution and superior image contrast. Usually found only on the most expensive models.

Barrel- used in the field of optics to mean some type of metal housing or holder for lenses.

Baffle- a mechanical barrier designed to lessen the effects of stray light on a lens system. In binoculars, may consist of finely machined ridges within the optical tubes.

BAK-4-A common designation for a high-density barium crown glass. BAK-4 makes a better porro prism because less light leaks out of the light path than in ordinary crowns. Lower scattering and brighter images result.

BK-7-A common designation for borosilicate crown glass. Popular as one element of an objective lens, but not that great when used in porro prisms.

Blind Spot- the point at which the optic nerve enters the back of the eye has no optical receptors and is not sensitive to light. Although this is a very small area, it may sometimes be noticed using optical instruments.

Binocular Vision-The visual ability of many animals to view a subject from two slightly differing points due to the anatomical placement of the eyes. The result is a feeling of "depth " in relation to the object being viewed.

Binocular- an optical device with enough eye relief and physical size so that both eyes look at a single view. Common in night vision equipment.

Brightness of Image- as the image is view through an instrument, the brilliance or luminance imparted to the image by all elements of the optical system. Unless it can be compared to a"standard" a rather meaningless term.

BP2-UV- a binocular prism material that appeared briefly in the early 1980's. The term means"binocular prism type 2 ultraviolet". Optical performance characteristics are identical to BAK-4 with multi-coating so the material was discontinued is hardly ever encountered today.

Cell- the machined metal portion of an optical instrument holding the lenses . Both the eyepiece and the objective lenses are mounted in a *cell.*

Chromatic Aberration- when you look through an optical instrument at a bright object, the image seem to be tinged with blue and yellow. Commonly called "color". The less refined the lens system, the more chromatic aberration.

Collimation- the alignment of the optical axes of the lenses of a binocular to the machined mechanical components.

Coma- an optical defect seen when a bright star is viewed near the edge of the field. The star image appears like a tiny "comet" with a long tail pointing toward the edge of the field.

Erecting System- A series of lenses or prisms placed in an optical system to make the image appear right side up and correct in right and left orientation.

Eyepiece- also know as an "ocular". The eyepiece is that optical component that is *nearest* the eye.

Eye Relief- the distance from the last lens in the eyepiece to the point where the eye can view the edge of the image most comfortably. Tends to be a little less with wide angle or high magnification instruments.

Field of View- The maximum angle of view that can be seen through a binocular or optical instrument.

Fixed Focus- a binocular design that has no means of eyepiece adjustment and forces the muscles of the eye to accommodate for the various viewing distances.

Focus- adjusting the eyepiece of an instrument so that the image will be clearly seen by the user.

High Contrast- a desirable quality in an optical instrument. If we view a piece of white paper next to a piece of black paper, a good instrument will maintain this optical relationship. A poor instrument will tone down both colors to a sort of washed-out gray appearance.

Multi-Coated- the latest state of the art in optical coatings. While old optical coatings may be only one layer thick, multi-coating may add up to seven layers on the lens surface. This advancement increases light transmission even more than conventional coating.

Magnification-the ability of an instrument to make an object appear larger.

Night Glass- any telescope or binocular having an exit pupil of 7 or more millimeters. Gives the night- adapted human eye the maximum amount of light usable.

Optical Cement- a clear cement used to bond two or more lens elements together. Canada balsam has been used for years although it has now been replaced by more advanced epoxies.

Optical Coating- a very thin layer of clear material applied to an optical surface. The result is reduced light reflection from that surface and improved image contrast.

Objective Lens- that lens in a binocular closest to the target. Lets in the light to form the primary image age which in turn is magnified by the eyepiece.

"P" Coating- Also called *phase* coating. Used on better quality roof- prism binoculars to bring the light transmission shift caused by the unusual prisms back into synchronization.

Prism- a glass element in an optical system that allows light to be bent or change directions.

Prism chamber- in a binocular, that light- proof chamber containing the prism cluster.

Prism cover- a thin metal cover placed over the roof of a prism to prevent light scatter.

Prism shelf- the adjustable metal plate upon which the prisms are mounted.

Resolution- the ability of an optical system to show two very closely placed point sources as individual entities. The *higher* the resolution, the better the binocular.

Reticle- also know as a cross hair or reticle scale. A mechanical or optical device placing a scale or measuring device in an optical system. Binoculars may have rangefinding scales built into the ocular.

Rubber Coated- a thin layer of cellular rubber applied to the external surface of instruments to give a slight degree of shock protection. Always a *disadvantage* when trying to service or repair an instrument.

Scratch- a defect on a lens surface caused by abrasive action. A very fine hairline scratch is known as a *sleek*.

Zoom Optics- by internally moving a series of lenses within an optical system, the focal length or power of the system is constantly variable.

Sources

TOOLS - LENS CEMENT

Edmund Scientific
Dept 15B- N937 Edscorp Bldg.,
Barrington, NJ 08007

(Ask for Industrial / Educational catalogue)

BINOCULAR REPAIR SERVICE

Redlich Optical
711 W. Broad St
Falls Church, VA 22046
(703-241-4077)

BINOCULAR HOLDING DEVICES/ OBSERVING CHAIRS

Astronomical Innovations
P.O. Box 1483
Lenexa, KS 66285

Vista Instrument Co.
P.O. Box 1919
Santa Maria CA. 93454

BINOCULAR CARRYING SYSTEMS

The Final Product Inc.
77 South End Plaza
New Milford CT 06776

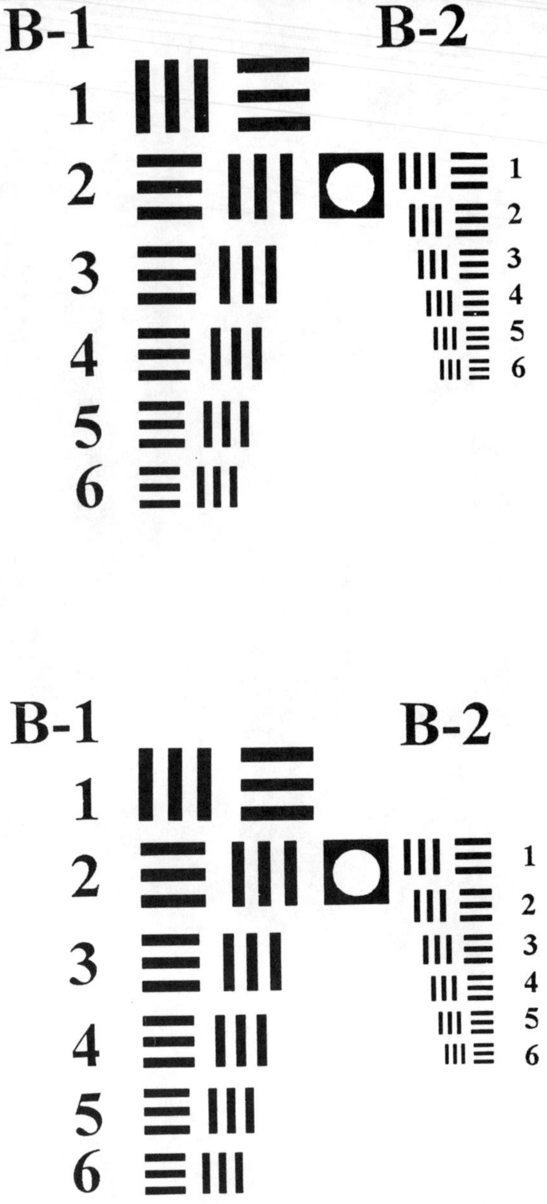

Binocular Test Chart

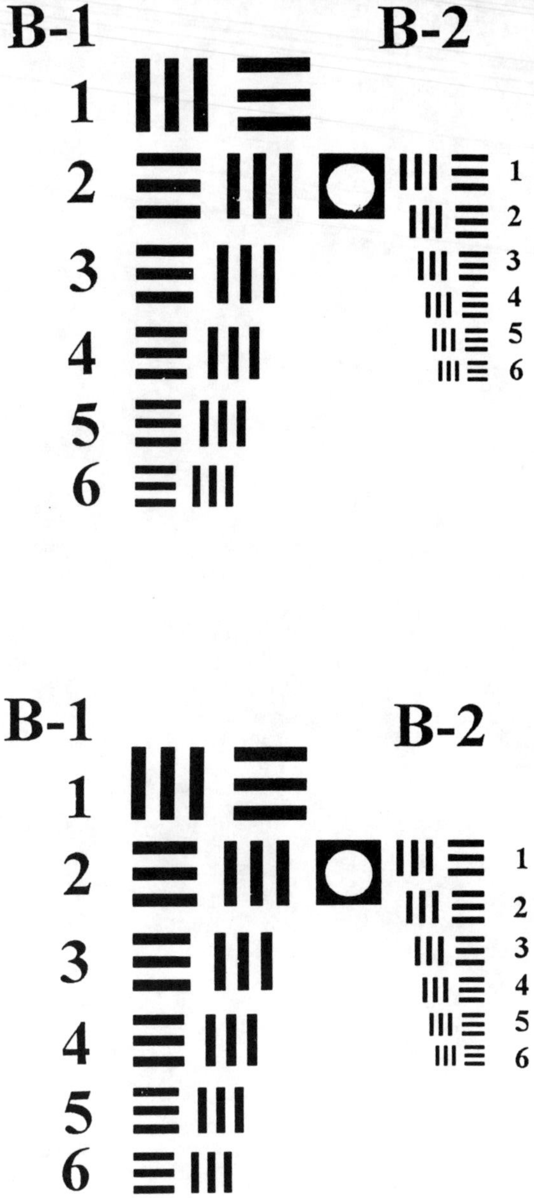

Binocular Test Chart

Bibliography

Brown, Sam., *All About Telescopes,* Edmund Scientific Co.,Barrington ,NJ. 1975

Bureau of Naval Personel. *Basic Optics and Optical Instruments,* Dover Publications, NY. 1969

Hale , Alan R. *How to Choose Binoculars,* C & A Publishing, Redondo Beach, CA. 1991

Ingalls, Albert G., *Amateur Telescope Making - Book Three* Kingsport Press, TN. 1961

Paul, Henry., *Binoculars and All Purpose Telescopes,* AMPHOTO, NY. 1980

Robinson, Leif J. *Outdoor Optics,* Lyons & Burford., NY, NY. 1989

Index

Notes

Notes

Notes

Order Form

Telephone Orders:

Call Toll Free:
1(800) 521-2828

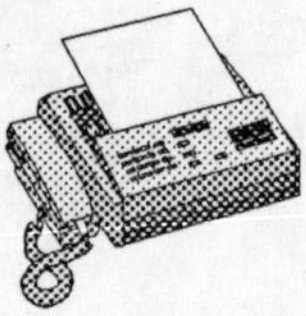

Fax Orders:

Call: 1(313) 665-1815

Postal Orders:

University Optics
P.O. Box 1205
Ann Arbor, MI 48106

Please send __________ copy (copies) of
Choosing, Using & Repairing Binoculars
@ $19.95 ea. for a total of $__________.
(Postage paid.)
Michigan Residents: Please include 6% sales tax..

Name: ____

Address: ____

City: ______________________ **State:** _______ **Zip:** ________

Payment:

______Check
______Credit Card: ☐ Visa ☐MasterCard ☐Discover

Visa-Mastercard-Discover
Account Number:

Expiration Date: ______________________
Signature: __________________________